The Northeast Passage

The
Northeast
Passage

BLACK WATER, WHITE ICE

Helen Orlob

THOMAS NELSON INC., PUBLISHERS
Nashville   New York

First edition

Library of Congress Cataloging in Publication Data

Orlob, Helen.
　The northeast passage.

　Bibliography: p. 121
　Includes index.
　SUMMARY:　Relates the attempts since 1553 of men from various nations to navigate the dangerous Northeast Passage through the Arctic to the Pacific Ocean.
　1.　Northeast Passage—Juvenile literature.
[1.　Northeast Passage]　I.　Title.
G680.074　　　910'.09'1632　　　77-9064
ISBN 0-8407-6564-9

For Dain and Brett

Contents

The Northeast Passage

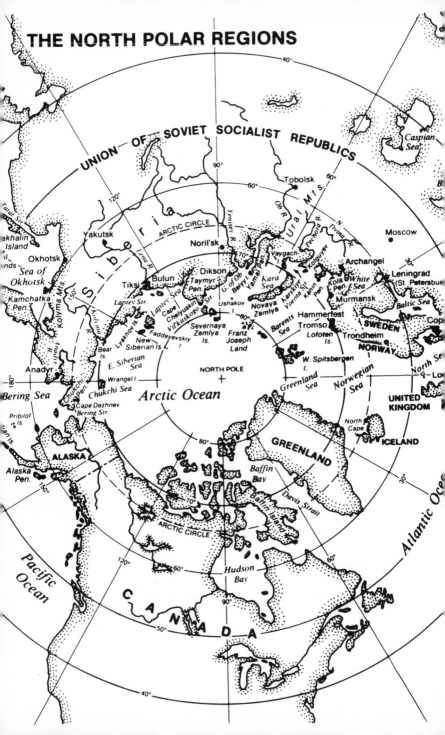

THE NORTH POLAR REGIONS

1 Where Strangers Are Not Welcome

On a gray afternoon in mid-July, 1965, there was brisk activity on Copenhagen's Langelinie Pier as the United States Coast Guard icebreaker *Northwind* prepared to sail for the Arctic after three days in port. Last-minute mail and supplies had come aboard, men on the pier stood ready to cast off, and most of the ship's crew lined the rails. A small crowd of the friendly Danes who had toured the vessel during its stay was on hand to shout a farewell.

Its white paint gleaming, flags snapping in the breeze, whistle sounding a salute, the *Northwind* started a slow movement away from the pier at 3 P.M. Cries floated up from dockside: "We'll look for you in the fall" and "See you when you come back this way."

Grinning and waving, the crew hoped to a man that it would not be so. They had another notion about where the ship's northern adventure might end.

Among the last to board the ship before the lines were let go was a young newsman, Richard Petrow, special correspondent to *The New York Times,* who had just filed a dispatch at the main post office in town. He was still fuming

11

with the frustration he had felt when he wrote that last story before sailing. Upon joining the *Northwind* in Copenhagen, he had thought that he would be sending an entirely different one. Then he had had the word from the icebreaker's commanding officer, Captain Kingdrel Ayers: "The State Department says there is to be no news about our intentions up there in the Arctic. That is, not until after we succeed—if we do."

What on earth were his editors going to think when they got that piece, after he had promised so much? "A three-month voyage of scientific exploration . . . into the Soviet Arctic," Petrow had been obliged to call the cruise planned for the *Northwind*. He had done his best with the story, adding details on the icebreaker's route, current conditions in the Arctic, and the scope of the studies to be undertaken, but he was bitter with disappointment.

When, if ever, he wondered, would the U. S. State Department permit him to break the news of the ship's real mission—a voyage uprecedented in history, the first American transit of the Northeast Passage from the Atlantic to the Pacific?

Certainly, the daring plan was no secret aboard ship. Everyone was aware that it called for the *Northwind* to sail through seas claimed by the Soviets as their own, and none knew better than Captain Ayers that his ship was to enter waters where strangers were not welcome. Consequently, he was not surprised when, a few days out of Copenhagen, a Soviet vessel started circling the *Northwind* while she was still off the coast of Norway.

With all hands alert for further inspection, the icebreaker passed North Cape and entered the Barents Sea. The expected watchdogging began almost at once. A Soviet bomber roared over, camera doors open. The plane cleared the *Northwind* with about one hundred feet to spare and disappeared. A couple of hours later another came snooping with two low passes. It was followed by a third which dropped a flare signal that was clearly understood to mean, "Attention. You are a violator."

Since the signal was not in aircraft-to-ship code, Captain Ayers did not feel obliged to acknowledge a reading. "Not understood," the *Northwind* replied. The overflights continued at two-hour intervals for the rest of the day, although there were no further signals.

The next day Soviet destroyer 020 came surging up to begin what developed into a quite amicable exchange of information:

020: "Where you should steer?"

Northwind: "Kara Sea."

020: "Where you from?"

Northwind: "New York. Where you from?"

The reply was not understood by the Americans, although it seemed probable that the destroyer had sailed from Murmansk, home port of the Soviet Northern Fleet and the largest naval base in the world.

020: "How long will you be in Kara Sea?"

Northwind: "Twenty days."

020: "What is after end of program in area Kara Sea?"

Northwind: "Barents Sea. [Captain Ayers must have had his fingers crossed at this point, since he had no intention of returning to the Barents Sea following the Kara Sea expedition.] Are you going to Kara Sea with us?"

020: "Of course I am going with you."

That promise was not kept, however. Soon after midnight the destroyer charged up from astern, lights blinking a farewell message: "I wish you a pleasant voyage, Yanks. You are too good guys." She was soon lost to sight over the horizon.

A short time later the *Northwind* encountered ice and bucked her way into the Kara Sea after passing the northern tip of Novaya Zemlya (New Land), an 500-mile-long island chain which resembles a slightly curved index finger pointing at the mainland shore. There the scientists aboard began the oceanographical studies that were the icebreaker's professed mission in coming to the Arctic.

However, the initial bout with ice had bent a propeller shaft, and the ship was ordered back to Newcastle-upon-Tyne, England, for repairs. On the way to drydock, she had not been long in the Barents Sea when radar announced the approach of a vessel. It turned out to be 020 in as friendly a mood as before.

Captain Ayers decided to try the limits of this unexpectedly cordial relationship with his escort. He invited the Russian skipper to dinner on the following day. "Thank you," the destroyer signaled, ". . . shall be very glad to meet if a situation favorable."

The *Northwind*'s crew set about polishing her to utter perfection, and in the galley, planning for the dinner included the best the ship had to offer. It was all wasted effort. The next day the Russian blinked his regrets: "I am very sorry but I cannot do it," followed, as he made off toward the south, by an explanation: "In Russia due for dinner." No doubt Moscow had forwarded the word that it did not care to have a Soviet naval captain dining aboard an American ship.

The repair work at Newcastle went slowly. September had come before the *Northwind* left the drydock, but again her course was northeast, destination the Kara Sea.

Another picketboat had been waiting for her to emerge

from the British port, and not many hours later she had more company. A warship—a guided-missile frigate, no less—appeared from the south, white water rolling off her bows. She whipped past the *Northwind*, showing a signal that none of the Americans had ever seen displayed at sea before: "Stop immediately."

Tight-lipped with rage, Captain Ayers ignored the demand. His ship was sailing international waters, a hundred miles off the Norwegian coast. How dared the Russian give him orders?

The frigate captain had atrocious manners. Sweeping wide, first to one side, then the other, he made repeated passes at the *Northwind,* as though to ram. Each time he cut close astern, and at last fell back to become a grim shadow for the rest of the day. Sometime during the night he turned away.

Without further harassment, the *Northwind* crossed the Barents Sea and again passed north of Novaya Zemlya to enter the Kara Sea. But as she resumed the track of oceanographic stations, gradually edging eastward toward Cape Chelyuskin, the most northerly point of the Eurasian land mass, the Russians took a keener interest than ever in her activities. Bombers appeared, some at high altitude, some swooping in at masthead level. The overflights became more frequent, and another Soviet destroyer put in an appearance.

In mid-September the story of the *Northwind*'s 1965 cruise neared its climax. With the destroyer knifing silently along in her wake, the icebreaker was approaching narrow Vil'kitskogo Strait, which separates the Siberian mainland at Cape Chelyuskin from an island group called Severnaya Zemlya (North Land).

The Russians had never been misled by the Americans' scientific routine in the Kara Sea. Moscow had learned about the planned transit not long after it had been ordered and was resolved to halt it by one means or another. What Moscow did not know (or perhaps would not credit) was that sharp diplomatic protests sent to Washington had had their effect. Captain Ayers had received notice of the cancellation of his original orders while the ship was in Newcastle.

In any event, the Soviet Union had no intention of leaving the door to the eastern seas of the passage unbarred. When the *Northwind* dropped her anchors thirty miles from the entrance to Vil'kitskogo Strait in the late afternoon of September 17, she remained under vigilant scrutiny from the destroyer's bridge. No one knows what might have happened if she had moved any closer to the strait, for there can be no doubt that the Soviet warship was there to stop her.

Thirty hours later Captain Ayers ordered the anchors up, and the *Northwind* slowly got underway on a northerly heading. The destroyer steamed off on another course, and the confrontation was at an end.

Richard Petrow had had an adventure, but he did not have the story he had hoped to report. The full account of the American attempt to transit the Northeast Passage in 1965 did not appear in print until his book, *Across the Top of Russia,* was published two years later.

2 A "Newe and Strange Navigation"

Returning to the Atlantic, the *Northwind* sailed an ancient route of defeat. For more than three hundred years men had pitted the frail timbers of sailing craft against the polar ice, seeking a way from one ocean to the other. Scores of ships had been lost and hundreds of lives before a Swedish explorer finally forged through the entire passage with an auxiliary steam whaler in the late nineteenth century.

The English had begun the search. In 1553, three ships rounded North Cape bearing men who expected a long voyage but no very great difficulty in finding a way to the Pacific by this "newe and strange navigation." They were relying upon the assurances of Sebastian Cabot, distinguished geographer, mapmaker, and explorer. History had provided enough information regarding the Arctic to convince Cabot that there must be a northeast water route. There was King Alfred's chronicle of the voyage of Othere, the Norseman, for instance.

A ninth-century sightseer, Othere had sailed north beyond the known limits of civilization simply to find out what was there. He passed North Cape, and after skirting the coast near

Murmansk, made his way as far as the mouth of the Dvina River on the White Sea before turning back. He did not mention trouble with ice or even seeing it.

In Cabot's own time there had been a story that seemed to prove the truth of Othere's observations. A merchant, Gregory Istoma, had told of taking passage on one of a number of boats bound for Trondheim, Norway, from the White Sea. During the voyage he had been terrified by stormy seas and the hazards of sailing along a rocky coast, but he, too, had said nothing about ice. And his story was clear evidence of an established commerce route between the White Sea and ports on the Scandinavian peninsula.

If it was possible to sail that far without trouble, Cabot reasoned, would not the way beyond be clear? The medieval concept regarded the earth's size as much smaller than it actually is, and Cabot was sure that the distance to be traveled from the White Sea to the Pacific was not great.

England had for many years been hungering mightily for a share of the rich trade with the Indies in which the Spanish and the Portuguese had engaged after Vasco da Gama's voyage of discovery around the Cape of Good Hope in 1497, and as the sixteenth century advanced, there was growing speculation among English merchants about the existence of that Arctic route to the wealth of Cathay of which Cabot had talked. In fact, its discovery was vital to their hopes, for the sea power of Spain and Portugal lay athwart the long and costly southeastern route.

At length, a group of London investors sought out the aged explorer (he was then in his seventies), and began to ''deale and consult diligently with him.'' When he assured them that a northeast passage through the Arctic to the Pacific probably existed, the Company of Merchant Adventurers was organized to finance an expedition. Cabot became its governor and supervised the building of three ships for the voyage.

According to the standards of the time, the vessels were marvels of construction. The hulls, made of "strong and well seasoned plankes," fitted with the greatest skill, were caulked, pitched, and, finally, lead-sheeted for protection against the marine worms that were said to infest the warmer seas they would eventually sail upon. They were provisioned for an eighteen-month trip.

Christened *Bona Esperanza, Bona Confidentia,* and *Edward Bonaventure,* they were ready to sail before the shareholders had finished debating the selection of officers for the expedition. Many had offered themselves for the command, some of them ill-qualified indeed, but the merits of each for the post of "Captain Generall" must be considered by the whole company. The review of candidates ended with the appointment of Sir Hugh Willoughby, "both by reason of his goodly personage . . . as also for his singular skill in the services of warre. . . ." (Historians have noted that nothing was said of his skills at sea.) Richard Chancelor, a man of exceptional qualifications, was named second in command of the expedition with the title of pilot major.

On May 10, 1553, the three ships left their moorage near London on the turning tide, and started down the Thames. Nearing Greenwich, where King Edward VI and his court were in residence, the sailors saw courtiers running from the palace. Others had crowded at its windows and were racing to the tops of the towers.

The ships' guns fired a resounding salute, while men manned the yards and tops, and the eleven Merchant Adventurers who had decided to go along waved from the decks. Only the pathetic fifteen-year-old king, too ill to rise from his bed, missed the pageantry of the expedition's departure. (Edward VI died of tuberculosis less than two months later.)

The little fleet made very slow progress during the first weeks of the voyage, but there was no serious trouble until it reached the Lofoten Islands, off the Norwegian coast, in early August. Willoughby intended to anchor overnight within the islands, from which a friendly native, who had rowed out to greet the strangers, promised to guide him. He knew the way around North Cape, he said, and as far beyond as a watchtower on the northern coast (Vardohus, site of the modern city of Vardø). Before the ships could reach the island haven, a violent storm drove them to sea under bare masts.

On the following morning, Willoughby, whose flag of command flew in the *Bona Esperanza,* searched the horizon for the other ships and at last made out the *Bona Confidentia.* The third ship was missing, nor did he ever see her again. Four days later, when the wind abated, he ordered sail hoisted, and the two ships continued to sail northeast. The commander was confident that he would find the *Edward Bonaventure* at the watchtower, which had earlier been designated a rendezvous point.

Probably because he had been driven far from the islands which would have supplied a pilot, Willoughby gave no further thought to such aid. He must have regretted the mischance of the storm more than once during the weeks that followed, for after rounding the cape, he could not find the tower. At the mercy of contrary winds, he sailed to and fro until the end of September, when he decided to winter on the Arctic coast, in order to continue the voyage the next year.

Willoughby's ships took refuge in a harbor far to the southeast of Vardohus. Scouting parties sent in different directions to seek the help of natives returned "without finding of people or any similitude of habitation," and the sixty-odd explorers, including nine of the merchants who had departed Greenwich in such high spirits, settled down to wait out a long night of cold.

Details of the wintering are lacking, for none of the party survived. During the next summer Russian fishermen came upon the ships and found the bodies.

Willoughby's journal and will made it clear that most of them had been alive in January and that the commander knew then that his own death was close at hand. He gave no particular cause for the tragedy that was overtaking the expedition, and it was said at the time that his people had died for lack of experience in "making caves and stoves." Certainly they did not perish of starvation, for well-preserved food was found with them, so it seems likely that scurvy, the disease caused by a diet deficient in Vitamin C, must be added to exposure to account for the disaster.

While the unfortunate men of Willoughby's command were dying, the lively pilot major had chanced upon quite another sort of fate. He was being magnificently entertained at the royal court in Moscow.

Richard Chancelor had seen the last of the two companion vessels at the outset of the storm on the Norwegian coast. He saw their sails taken in and the *Bona Esperanza*'s pinnace swept overboard to be smashed against her side—after that, nothing. When the gale died down, he too sailed on northeast, and being an experienced seaman (which the captain general was not), he had no trouble finding the place of rendezvous.

There he waited for seven days before resolving to go on "either to bring that to passe which was intended, or else die the death. . . ."

The *Edward Bonaventure* sailed southeast, following the coast, as Othere had done, and eventually entered the White Sea. At the mouth of the Dvina, near a monastery, Chancelor had the anchor dropped and a boat lowered in order to go in pursuit of a fishing craft sighted in the distance. The fishermen did their best to escape, but the Englishmen succeeded in boarding the craft, whereupon the terrified natives flung themselves down, trying to kiss Chancelor's feet. With comforting signs and gestures, he pulled them up, making it plain that the great ship which had so frightened them had come in friendship.

The news of the coming of strange men who were all gentleness and courtesy raced through the populace of the riverbanks. Shortly, Chancelor was playing host to crowds of curious people who came with gifts of food. They fingered the goods that the merchants offered in exchange for their products, and there was no doubt that they coveted much of it, but trade they would not. By one means or another, they made it clear that such traffic must have the approval of their king, the ruler of a vast land called Russia or Muscovy, and he dwelt many, many miles to the south.

Secretly, a messenger had already been sent posthaste to Moscow to learn the pleasure of Ivan IV (Ivan the Terrible) regarding the strangers, and though Chancelor pressed the natives to forget their scruples about trade, they continued to refuse in the absence of a return message. After several weeks, he threatened to sail away with his treasures of cloth and metal, and that brought an offer to send him to Moscow, which he accepted readily.

Ivan's northern subjects had really anticipated the czar's own wishes, for midway on the long sled journey that Chancelor and a few companions undertook, they met the imperial messenger, who bore not only permission to trade but a hearty invitation to visit the court.

Chancelor remained in Moscow for three months, time enough to learn a little of the language and to observe the

Russian way of life. In general, he was unimpressed. He thought the Russian capital must be as big as London, but for "beautie and fairnesse, nothing comparable. . . ." With some superiority, he noted rudely furnished houses built of wood lining unpaved streets. Even the king's palace was "much surpassed by the beautie and elegancie of the houses of the kings of England."

The tone of his account alters abruptly, however, with the description of Ivan and his court. The astonishing quantities of gold, silver, and jewels that he saw left him in no doubt that commerce with Russia would be profitable for his countrymen. He hastened to exert all his charm to promote it, and his success is proved by the agreements he carried with him when the *Edward Bonaventure* sailed for home the next summer. They led to the establishment of a long-used trade route between the White Sea and England.

Thus, the Willoughby expedition ended with a measure of triumph, as well as tragedy.

Richard Chancelor did not live to know the extent of either. He left England for another trip to the White Sea and Moscow in 1555 before news of the discovery of the two missing ships had been received, and returning in 1556, lost his life in a shipwreck on the Scottish coast.

3 The Defeat of the Merchant Adventurers

The Merchant Adventurers were delighted with the results of Chancelor's winter in Moscow and prepared to take immediate advantage of a trade, which, if less exotic than the one they had hoped for, could not fail to be profitable. Each nation stood in need of what the other had to offer. England had woolen cloth, arms, and other metal products to exchange for Russian furs, tallow, hemp, and train (whale) oil.

Nevertheless, Cabot and the merchants still dreamed of spices, precious metals, and jewels, and Chancelor's observations on the opulence of Ivan's court led them to believe that the way from Moscow to the East must be short. Surely it would be found just around the corner from the river mouth where the *Edward Bonaventure* had anchored. Even as Chancelor sailed in 1555 on the first voyage of the new trade, plans were being made to send out a single ship whose success in finding the route could not be doubted.

Stephen Burrough, a young man of spirit, who had been master of the *Edward* on the first voyage, was named to command the *Searchthrift,* a tiny vessel which sailed down the Thames in the spring of 1556. He logged North Cape a month after leaving the English coast, and a week or so later anchored his ship with a number of Russian *lodjas* (small clinker-built vessels supplied with sails and twenty oars) in a river mouth on the coast near Murmansk.

Burrough was not slow to realize that he had had a stroke of luck in coming upon the Russian seamen. Pheodor, master of one of the vessels, came aboard to present "a great loaf of bread, and six rings of bread . . . and four dried pikes, and a peck of fine oatmeal," for which he received rather stingy thanks—a comb and a small glass. However, when Pheodor's signs made it clear that the Russians were going east to the Pechora River, Burrough hastened to add refreshments to his small gifts, for that was exactly where he wanted to go, and nothing would please him more than to tag along in the wake of an experienced pilot.

For seven days, while other *lodjas* continued to appear from the river, the Englishman waited for the fleet to move. At last, impatient with such dawdling, he weighed anchor and put to sea. The Russians watched the departure, probably with much headshaking, and could not have been surprised to see the vessel reappear after several hours. Outside the anchorage, Burrough had found vicious seas and a wind with which he could not contend.

The *Searchthrift* had visitors aplenty that day, all bearing gifts and advice. Among them was Gabriel, who was to prove himself a real friend to Burrough. Somehow the Russian made him understand that with a fair wind the voyage to the Pechora would be made in seven or eight days. If Burrough would follow him, he would warn him of shoal water.

Gabriel sealed the pact of friendship by delivering to the *Searchthrift* a barrel of mead, and another *lodja* captain provided a barrel of beer. The next few days saw a good deal of international partying, during which the Russians drew maps to describe a way east through an island chain called Novaya Zemlya and as far beyond as the Ob' River.

When a favorable wind came at last on the second day of July, the *Searchthrift* hoisted sail in company with thirty *lodjas*, but Burrough was shortly much embarrassed to find that he could not keep up with the Russians. Gabriel and one of his friends struck their sails often, waiting for the laggard to catch up. Burrough had good reason to be thankful for the help. According to his account, the course lay over "sunke land, and . . . you shall have scant two fadome (twelve feet) water and see no land."

Gabriel continued to befriend the English captain during a wild passage to the Pechora River. Several times he guided the *Searchthrift* to a safe refuge from storms that made his earlier prediction of a seven-or eight-day voyage a sad joke. Once, he helped free the vessel when it grounded on a sandbar, and he took Burrough's part in a quarrel concerning a hawser that one of the Russians was trying to pilfer. No doubt the two men parted with mutual regret at the river mouth, when the *Searchthrift* made ready to sail on alone into the northeast.

Burrough remembered the week that followed as one filled with the terrors of the far northern latitudes he was reaching. On the day after leaving the Pechora, he thought that he had land in sight and made for it, only to discover "a monstrous heap of ice," the first that any explorer had encountered. A half hour later the ship was surrounded by drifting floes, and he worked her this way and that, dodging disaster.

The next two days brought more ice from which he escaped to face another awful sight: "a monstrous whale aboord of us, so neere to our side that we might have thrust a sworde or any other weapon in him, which we durst not doe for feare he should have overthrowen our shippe.'' Burrough called the crew together to shout in concert, at which the whale disappeared with a fearful commotion and noise, and they all thanked God for deliverance. Later that day they were more than happy to reach a safe anchorage among some islands.

Burrough did not doubt that these were the islands, described by his late companions, which barred the way to the sea into which the Ob' poured its waters, and while he quested about, seeking the passage of which they had told him, he came upon Loshak, another friendly Russian. Yes, indeed this was Novaya Zemlya, said Loshak, but Burrough had missed the strait he sought. Upon receiving several gifts, he described in considerable detail the proper course: south along the shore of Vaygach Island to a narrow passage between it and the mainland.

Storms and drifting ice kept the *Searchthrift* at anchor much of the time during the first half of August, and again Burrough was in the company of Russians. He watched them chase a white bear over the cliffs, in order to kill it in the sea, and with them, went ashore on Vaygach to see wooden idols—crude representations of men and women, which the native Samoyeds worshiped. The people of this country were not much to be feared, he was told, but in the Ob' country, it was another matter. There ''they will shoot at all men to the uttermost of their power, that cannot speake their speech.'' The Russians went to the Ob' only when the hunting here in Novaya Zemlya failed—as it had this year, they added sourly.

The walruses were, in fact, so scarce that Loshak at last told Burrough that he had made up his mind to go through the strait and cross the sea beyond (the Kara Sea) to the Ob'. The Englishman might follow him, if he cared to.

Loshak may have made his bid for better hunting, but it was too late for Burrough. The days were growing much shorter, the storms were increasing in severity, bringing with them great quantities of ice. On the first day of September he decided to abandon exploring for the year and sailed west to winter in the White Sea. He planned to come back in the next summer to bring his quest to a successful conclusion.

Gabriel, Loshak, and the other Russian friends never saw Burrough again. In 1557, he was asked to sail along the White Sea trade route in search of three vessels that had been lost, and when he reached England after a fruitless voyage, he found that interest in the search for an Arctic passage to the Indies had ended with the death of Sebastian Cabot.

The Merchant Adventurers, now calling themselves the Muscovy Company, were trying, instead, to establish a route to the East by way of Moscow, the Volga River, and the Caspian Sea.

Still, there were learned men and navigators of great repute who continued to point to the north as the only practical way to Cathay. Among them, Martin Frobisher talked for years of a northwest route, ''as plausible as the English Channel.'' Eventually, he made three unsuccessful voyages in search of it. Charles Jackman, one of his company, thought the northeast a more promising direction. So, too, did his friend, Arthur Pet, who had sailed with Richard Chancelor as a youth. The two Arctic veterans at last contrived to bring their notions to the attention of the Muscovy Company so persuasively that the Merchant Adventurers decided to risk one more northeastern voyage.

Sailing from England in the spring of 1580 in two tiny ships, the *William* and the *George,* Pet and Jackman separated shortly after rounding North Cape, agreeing to meet at Vaygach Island. Both captains succeeded in taking their vessels beyond Novaya Zemlya into the Kara Sea, the first explorers to do so, but the masses of ice encountered there forced them to turn back in August.

Once again in England, Pet told the story of the voyage alone, for Jackman's *William* had been lost with all aboard on the way back.

His account, read in Dutch translation more eagerly than at home, was to provide the impetus for the next adventures in northeast exploration.

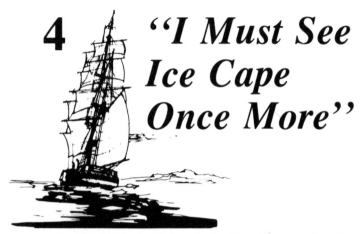

4　"I Must See Ice Cape Once More"

Across the North Sea in Holland, the Dutch, struggling for independence from Spain and stirring with mercantile ambition, had heard accounts of the first English voyages to the Northeast with more than casual interest. Scarcely twenty years after Chancelor's voyage they had taken advantage of his discovery by establishing a White Sea trade with Russia, and when a translation of Arthur Pet's journal came to their hands, it was no more than confirmation of what they already knew. In fact, a Dutchman, Oliver Brunel, had seen more of Russia than any Englishman had.

Brunel, who deserves better of history than the few scraps of information we have about him, was an agent at Holland's trading post on the Dvina River. During his years in Arctic Russia he traveled extensively, once reaching the mouth of the Ob' River on an overland journey. He was told that Russian vessels regularly made the run from the White Sea past the Ob' to the Yenisey River, where they wintered, and that beyond the Yenisey the way was open to the east. Four years after the Pet–Jackman expedition, Brunel tried to find that route in a ship laden with trade goods. He was turned back before he had gone as far as the English had, and subsequently lost his vessel on shoals at the mouth of the Pechora River.

Ten years after the Brunel disaster, the Dutch elected to try again where the English and their own man had failed. They did not doubt that the passage existed. They had immense faith in the abilities of their navigators and, newly independent, were more than ever eager for a share of the wealth of Cathay.

The expedition which sailed from Holland in the spring of 1594 was made up of four vessels—two named *Mercurius,* one under command of Willem Barents, the other under Brandt Tetgales, the *Swan,* captained by Cornelius Nay, and a small fishing sloop, which was to be a companion vessel to Barents' *Mercurius.*

After reaching the Arctic Ocean, the ships separated. Nay and Tetgales, under orders to follow Arthur Pet's sailing directions, set course for Vaygach Island, while Barents made his way north along the coast of Novaya Zemlya in an attempt to round its northern extremity into the Kara Sea.

There were hazards aplenty in the unknown waters in which Barents' *Mercurius* and her consort sailed. Some of them were completely unexpected—polar bears, for instance. The Dutchmen stared at the first one sighted, a great white creature swimming in a sea adrift with ice, and joked so much about the sport of taking the fellow home to Holland alive that a few of them, including the captain, finally launched a boat to go in pursuit. They managed to toss a noose around the beast's neck and started towing it back to the ship. The lassoed bear soon had enough of that and swam up to put a forepaw over the gunwale, as though "to rest himself a little," said Barents. The captain was mistaken. The captive was going after his tormentors, and was only put away after he had heaved himself half into the boat. The Dutchmen were thankful enough to be returning with only the skin of such a trophy.

During the latter part of July, as the *Mercurius* reached higher latitudes than any ship had yet logged, Barents fought ice for every mile of northward progress. At last there came a day when no movement was possible. The explorer named the promontory sighted to the east Ice Cape and reckoned his position as 77 degrees north. He had reached the tip of Novaya Zemlya, but there was no way past it.

Two weeks of waiting brought no break in the ice field, and the lateness of the season, together with his crew's apprehensions, obliged Barents to give up. He turned away sadly, sure that this was the best way into the Kara Sea. Late in August he kept a rendezvous with the rest of the fleet near the western shore of Vaygach Island, and heard from Nay and Tetgales a jubilant tale of having sailed as far as the mouth of the Ob' River, from which the remainder of the passage to Cathay would be easy.

The two explorers had not, in fact, been near the Ob'. After parting from Barents, they had reached Vaygach, where Samoyeds gave them remarkably accurate information about sailing beyond Novaya Zemlya. Scorning the words of the natives, who could know nothing of navigable waters, Nay and Tetgales went on through the strait and sailed into the Kara Sea. Several days later the captains dropped their anchors off the mouth of a river which they were sure must be the Ob'. The misidentified river was probably the Kara, and at this point, on August 21, they turned back.

Following the triumphant return of the 1594 expedition, the Dutch made enthusiastic plans for a fleet of seven vessels to sail to Cathay the next year. Little time was left in which to prepare for so ambitious an undertaking, and as a result, the summer of 1595 was well advanced before the fleet sailed. Nay was in overall command, and again Barents was a ship captain.

Nay must have been a bit dubious about success from the beginning, for Vaygach was not sighted until the end of August, the very time at which the previous year's expedition had started for home, and there was much more ice than had been seen on that voyage. The ships did succeed in passing through to the Kara Sea, but massed floes forced a halt at an island only a short distance from the strait.

None of the captains had much stomach for further effort that year, and when the horror of seeing two men of a shore party mauled to death by a polar bear was added to their discouragement, they agreed to turn back.

For a time it seemed that the 1595 fiasco might mark the end of Barents' adventuring in the Arctic. The Dutch government, which had provided the funds for the first two expeditions, was not inclined to make further investments of the sort. However, it offered a prize for any vessel that might succeed in making its way through a northeast passage, and two groups of Amsterdam merchants decided to make a bid for it in 1597.

The pair of ships that sailed from Holland in May of that year were commanded by Willem Barents and Jan Cornelius Rijp. Late in June they came upon the western shores of Spitsbergen. Ice made further progress to the north impossible. In retreat, the two captains argued about the future course. Rijp wished to try the eastern coast of this newly discovered island group for a passage, while Barents insisted that the only possible way was much farther east. Hot words flew at a final meeting, and Barents set off alone for the place that he had named Ice Cape three years earlier.

Conditions were better than in 1594, and at the end of August the Dutchman became the first explorer to pass the northern extremity of Novaya Zemlya. His jubilation did not last long, for try as he would, he could find no way farther east. Solid masses of ice confronted him in every quarter.

"We must return to Holland," he decided after five days of frustration, but he was too late. Overnight the waters of the shallow bay where he had anchored turned to ice. The ship was fast, and there was no hope that September weather would free her. Barents and his men would have to winter in a region where it was not supposed that humans could survive the bitter cold of the dark months.

The Dutchmen, sixteen men and the ship's boy, faced the prospect with courage and energy. The ship could not be thought of as a winter refuge. They must have a house ashore. Scouring the vessel for planks and other construction materials, they set to work, and by the end of October they had built a snug, tight dwelling some twenty by thirty feet in size. Around it they heaped great piles of driftwood to burn in the central fireplace, and they carried ashore every article that promised to be useful for winter existence—even a large barrel to serve as a bathtub.

Not long after they had moved into the house the sun went down for the last time in 1597. The long night had come, and with it, cold of an intensity none of them had ever imagined. They saw its effects with awe—clothing that stiffened like a board as soon as it was lifted from warm water, socks that burned off their feet before the heat of the fire had been felt, beer that turned to a gluelike substance.

One storm followed another with scarcely a lull between. Snow piled deep around the house. At times, it was necessary to use the chimney for an exit—although not many cared to leave the four walls of the refuge during a blizzard.

During the building of the house, polar bears had prowled the shore, and a few of them had supplied meat for the cooking pots. They had disappeared with the sun, to be succeeded by foxes, who could often be heard padding about on the roof. An occasional meal of fox meat made a welcome addition to a diet that was barely adequate. Midwinter found many of the men suffering the first effects of scurvy, and when the sun appeared again in February, they were all in weakened condition. One man died that month, and Barents became very ill shortly afterward.

He had hoped to get away in May, but the ship was still fast at the end of the month. "Very well, we'll use the ship's boats," he decided. Sick men manned the two little craft when they set out on the long voyage home, and the captain was carried aboard. Six days later, again off the tip of Novaya Zemlya, Barents knew that death was near. "Lift me up," he told one of his companions. "I must see Ice Cape once more." A few moments later he died. His body was committed to the waters that came to be known as the Barents Sea.

Another man died that day, and a third about three weeks later. The rest of the crew survived a terrible voyage to the coast of the Kola Peninsula, where they were saved by coming upon Rijp's vessel.

Their story of living through the Arctic night, buried in history, came to light again 273 years later. In 1871, a Norwegian hunter, Elling Carlsen, was sailing along the coast of Novaya Zemlya when he saw on the shore of a small bay the ruins of a house. Ashore, he found that the roof had fallen in, but the walls, one and a half feet thick, still stood. He started shoveling at the masses of ice and gravel that filled the interior, came upon the fireplace with its spacing bar still in place, and found things which the Dutchmen could not carry with them when they abandoned the house that spring day in 1598—copper pans, a flute, the tattered remnants of books, a boy's shoe with the lacing still intact.

Later exploration revealed many more objects that had survived almost three centuries of deep freeze in the Arctic. Eventually, they were all turned over to the Dutch government, which displays them at The Hague in a museum room which is an exact replica of the house at Barents' Ice Haven, site of the first successful wintering in the far Polar North.

5 The Age of Tall Tales

As the sixteenth century came to a close, both England and Holland could count only defeat and tragedy as the rewards of nearly fifty years of searching for a northern passage to Cathay. It was a bitter reckoning, for the need for this route was not critical after England had sent the ships of Spain's mighty armada to grief on the rocks of the Cornish coast in 1588. The Spanish stranglehold on the Atlantic had been broken, and fleets of Indiamen, flying Dutch and English ensigns, had begun to use the southern routes.

Nevertheless, merchants on both sides of the North Sea still toyed with the notion of that shorter, more profitable transit of which they had dreamed. Did it truly exist? "Yes," said geographers. They were as sure of it as ever old Sebastian Cabot had been in the days of their fathers and grandfathers. But arguments raged when the matter of direction came up. Did the route lie to the northwest or to the northeast? Or perhaps due north?

There was no one to say about this last speculation, "Ridiculous! You can't sail across the North Pole." In fact, until comparatively recent times there were men who believed that if a ship could forge its way through a belt of Arctic ice, warmer water would be found leading straight to the Pole. Apparently, some of the merchants of the Muscovy Company were of this mind, for in 1607 they again decided to risk a little capital on a voyage of exploration in the north.

On a spring Sunday in that year, eleven men and a boy, "purposing to goe to sea foure days after, for to discover a passage by the North Pole to Japan and China," entered a London church to ask for divine blessing on their venture. The captain was Henry Hudson, the boy, his son John.

This is thought to be the first historical reference to one of the most famous explorers of all time. Hudson may have been a grandson of one of the original Merchant Adventurers, and it is believed that he had grown up in the service of the Muscovy Company. It seems clear that he had spent his life at sea, for he was a skilled seaman and navigator.

The 1607 voyage, on which Hudson sailed even farther north than Barents, failed of its goal, of course, as did one made the following year. But he brought his ship and crew home safely each time and was able to supply new information on the coasts of Spitsbergen and Novaya Zemlya.

Mutual disappointment may have led to bitter words between merchants and captain, for their ways certainly parted after the second voyage. Early in 1609, Hudson was in Amsterdam, seeking employment with the Dutch East India Company. In fact, he had been summoned to a quiz on his northern voyages. The company had been told that he had sailed to 80 degrees north and beyond. If that was so, what did he think of his chances of finding a passage? Hudson thought they were good, and again the Dutch began to look northward with hope.

The company was not willing to go ahead with an expedition that year, however, so the Englishman was paid for his trouble and dismissed with a promise for 1610. At once he began to dicker with the French, whereupon his late hosts promptly changed their minds and drew up a contract for his approval. Its terms were exact. About the first of April he was to sail in search of a northern passage, but he was admonished to "think of no other route or passages except the route around the north and northeast above Nova Zembla [Novaya Zemlya]." If he was unsuccessful, another route would be considered for a future voyage, the agreement promised.

The language of the instructions, together with the fact that Hudson was not to be paid if he failed to follow them, give rise to the suspicion that he had been making a pretty strong case for exploration in another direction, and that the East India Company was having none of it. They wanted to know more about that route of which Barents had been so sure and probably waved aside the words of their Englishman's friend, Captain John Smith of the colony at Jamestown, Virginia, who had written him of a possible passage through the North American continent.

Hudson sailed from Holland in the little *Half Moon* during the first week of April, 1609, and, passing North Cape a month later, set course for Novaya Zemlya. After two weeks of battling through the Barents Sea under the worst possible conditions, the *Half Moon*'s crew, a raffish lot, threatened mutiny. "Turn back, or we take the ship," they told the captain.

Hudson seems to have given in without an argument, perhaps because the malcontents were giving him a chance to do precisely what he had wanted. He informed them that Captain Smith had told him of a sea passage through North America in about the latitude of 40 degrees. Would they agree to a transatlantic voyage in that direction, or would they prefer to search to the northwest in the neighborhood of Davis Strait? The crew, having had enough ice and snow, lost no time in choosing Hudson's first proposition, and the *Half Moon* came about to seek the warmer seas of the western Atlantic.

The rest of Henry Hudson's story has nothing to do with the Northeast Passage, but it must be noted that, in turning his back on that part of the world, he set course for the exploration in the New World on Holland's behalf which had so profound an effect on American colonization.

For many years after Hudson's failure in the northeast, there was no interest in a further search for the passage, but in that period there was no lack of travel to the seas the early explorers had sailed, or to the shores they had visited. Whaling and walrus-hunting expeditions increased in number each season, and their leaders, seeking richer and yet richer grounds, grew more daring about the high latitudes. In fact, one Dutch skipper may have sailed along the coast of Franz Josef Land two hundred years before it was officially discovered. The spare language of hunters' log books was thus adding steadily to European knowledge of the Russian Arctic.

In contrast, the seventeenth century was also a time of tall tales told by romancers who had little to fear in the way of contradiction on this strange subject matter. One author-explorer, whose work had a wide circulation, claimed that three knots of wind might be purchased from Lapp magicians on the northwest coast of Norway. The first knot released would produce a soft breeze, the second a gale of wind, but beware that third, he warned. It bore a storm that could wreck a ship.

Seamen who had returned from northern hunting expeditions whiled away their leisure hours in regaling coffeehouse companions with wild stories of having sailed in open seas close to the North Pole. One Dutch sailor in Amsterdam said, in the hearing of an Englishman, that his ship had sailed two degrees beyond it. "Is this the truth?" he was asked.

"My ship is in Amsterdam. The others aboard will tell you it is so," he replied.

"And did you see no land or islands about the Pole?"

"No, and we saw no ice."

"What was the weather there?"

"Why, it was fine warm weather such as here in Amsterdam in the summer time and as hot."

A pamphlet written in England in 1674 by the King's hydrographer contained the last story. Accounts of high latitudes reached and open seas at the Pole were included in the Philosophical Transactions of the Royal Society. These were papers that commanded respectful attention, and their contents were generally accepted as the truth.

Among the most gullible of their readers were Wood and Flawes, a pair of captains with no knowledge whatever of sailing in ice. Apparently they swallowed all the stories whole and were so eloquent about the probability of success on a northeast run to China that Charles II was persuaded to send them forth in 1676.

The two-ship expedition was an ignominious failure. It is said that Wood, sailing in the *Speedwell,* was completely unnerved by his first brush with drifting ice, and when the vessel came to grief on the coast of Novaya Zemlya, he confessed himself helpless with despair: "All I could do in this exigency was to let the brandy-bottle go round, which kept them [the crew] allways fox'd, till the 8th July Captain Flawes came so seasonably to our relief."

After the party had returned to England in Flawes' little *Prosperous,* Wood was bitter in his denunciation of all the northeastern explorers—Barents and others whose reports had been reliable, as well as the romancers. So voluble was he about the absolute impossibility of sailing in the Arctic that the sorry venture put a period to northeastern exploration for almost two centuries.

6 "What Is Siberia?"

At the beginning of the eighteenth century, Arctic Russia was no longer a land of mystery. Hunters had established routes to the best of the hunting grounds, and merchants had followed in their wake, opening trading posts at strategic points on remote rivers. The Arctic coastline had been plotted with fair accuracy, and the seas north of Russia had been sailed routinely by fishermen for at least two centuries.

But Arctic Siberia was still no more than a name lettered across a blank space on maps of the Eurasian continent. True, adventurous Cossacks had begun the conquest of the vast territory almost a century before, but the sketchy news of their progress eastward from one great river system to another was slow in reaching Europe, and many of the reports concerning the wealth of the new land were disbelieved.

The loot of conquest finally made it clear that the Cossacks were opening up a treasure house. Furs of a beauty never before seen in Europe—the pelts of sable, otter, fox and seal—spilled from huge bales in St. Petersburg (Leningrad) and Moscow, and there were boxes and bags of walrus ivory.

At last, in the early years of the eighteenth century, Peter the Great decided that he must know the extent of this rich land that was being claimed for his rule. He ordered planning for one of the most ambitious projects of exploration ever undertaken by a nation. It was to be called the Great Northern Expedition.

There were questions aplenty for which Peter wanted answers. What were Siberia's dimensions and its geographical features? Who were its people? How far toward the North Pole did the Siberian coastline extend? What was its length?

But most of all, the czar wanted to know whether there was truly a land bridge between Asia and America.

For Peter there were to be no answers. He was on his deathbed when, early in 1725, he scrawled out final orders for the initial venture of the Great Northern Expedition, a voyage from far-distant Kamchatka on the Pacific Ocean to determine if the Old and New Worlds were linked. "Build in Kamchatka . . . one or two decked boats," the czar ordered the man selected for command. "Sail on these boats along the shore which bears northerly and which (since its limits are unknown) seems to be a part of America. Determine where it joins with America . . . draw up a chart and come back here." Peter also wished to know about American settlements that might be discovered, he said.

Vitus Bering, Dane by birth, senior captain in the Russian Navy after long service, was not deceived by the simplicity of his orders. The portly little blond commander knew that he faced a task of awesome proportions. Four thousand miles separated St. Petersburg from the place where the ships' keels would be laid, and who knew what sort of travel it would be? Moreover, every scrap of material needed for the building, save the timber, must be transported over those miles. When he set out from the capital in February, three days before the czar's death, twenty-five wagons loaded with anchors, cannon, cordage, sails, and other supplies had already been dispatched.

The rigors of the journey were beyond Bering's wildest expectations. The first twelve hundred miles to Tobolsk, the Siberian capital, were easy enough, but after that there was nothing but wilderness, and it became a matter of using the water courses wherever possible, building freight barges and abandoning them on each. There were mountains to cross and forests that no man had ever seen. On the steppes, horses and men wallowed almost helplessly in wide stretches of bog. Occasionally, the explorers came upon native villages, but in one eight-hundred-mile stretch, they found not a single inhabitant.

Season followed season as the months of 1725 and 1726 were marked off on calendars. Bering and his party did not reach the Pacific shore at Okhotsk until the spring of 1727, and the broad expanse of the Sea of Okhotsk still separated them from Kamchatka.

At this point months of fearful toil could have been avoided, had it been known that Kamchatka was a peninsula. Bering could have sailed on his voyage of exploration from the place where he first discovered open water. But the Sea of Okhotsk was believed to be an inland sea, so he set his people to work building a vessel designed to carry the expedition across to Kamchatka. Midsummer had come before the *Fortuna* was finished, and by the time the ferrying operation had been completed, snow was falling.

In the frightful conditions of a Kamchatkan winter, the expedition again took to rivers and trails—three hundred miles up one river to a crossing of the mountains, then three hundred miles down another to a tiny settlement on the Pacific Ocean.

The fourth spring since the departure from St. Petersburg was at hand when work was begun on the expedition's last vessel, the *St. Gabriel*. She was finished, launched, and rigged in a little more than three months.

Late in July, 1728, Bering sailed his ship out of the mouth of the Kamchatka River, at last underway on the voyage of exploration to determine whether Asia and America were linked by land. Some three weeks later the *St. Gabriel* lay off the coast of the Chukchi Peninsula with nothing but open water in sight ahead. Bering and his officers were arguing about how much farther north the land bridge could be when men who could supply an answer, if they would, appeared out of the fog. Eight natives had come out in a skin boat to investigate the strange ship. The Chukchis rested on their paddles at a safe distance, simply staring. There was no response to the hallooing and waving that urged them closer until, after a time, one of the paddlers leaped into the sea and swam to the ship's side.

He would not come aboard, but he would talk. The ship's interpreter, a native from northern Kamchatka, gabbled with him at length, while Bering fidgeted for the information that he guessed must be forthcoming from all the pointing this way and that. "Well?" he barked when the interview had ended. "The Chukchi told me the coast goes north only a little way," said the interpreter. "Then it turns west. He says also that there is a big island only a little way east of us here."

The *St. Gabriel* sailed eastward to a glimpse of the fog-shrouded shores of the island, which Bering named St. Lawrence in honor of the saint's day of its discovery. Then her prow was again pointed north. A few days later, the

Chukchi's description of the coastline was verified when the land bore off suddenly to the west. "There is no link between Asia and America," Bering claimed and pointed to the cape on their left. "This is the most northeasterly point of Siberia. We shall call it East Cape. It is time to turn back."

His officers did not agree with him. Lieutenant Spanberg, second in command and a fellow Dane, thought that they had not yet proved anything. They should go farther north. Chirikov, the young Russian lieutenant who was third in command, wanted to follow the coast west and winter when they were halted by ice. The ship had been provisioned for a year, no risk was involved, he insisted.

Bering protested the danger of the last suggestion. This was an inhospitable shore, with neither wood nor shelter available. The ship might be crushed in the ice, in which case none of her people would survive. He gave in finally to Spanberg's urging. They would go north another two days before turning back.

No land was in sight after forty-eight hours, of course. "You see," said Bering, "there is no land bridge. We have fulfilled our mission and must return to Kamchatka while we still can."

Again in Bering Strait, the explorers sighted an island which the commander named St. Diomede (now called Big Diomede). He did not suspect that fog hid another close by (Little Diomede), and set course for Kamchatka, ignorant of the fact that the American continent, less than thirty miles away, would have been visible on a clear day.

Bering had no idea that he was not the first to sail through the strait between the Asian and American mainlands. Not long after the announcement of his discovery, old records revealed that in 1648 a Russian hunter, Semen Dezhnev, had passed through it on a voyage from the Kolyma River, which flows into the Arctic Ocean, to the mouth of the Anadyr River on the Pacific Ocean.

About 170 years after Bering sighted the northeast tip of Siberia, Dezhnev's feat gained belated recognition when East Cape was renamed Cape Dezhnev.

7 Captain Commander Bering

During the winter in the Kamchatkan harbor, Bering worked on his charts and reports, and he had the *St. Gabriel* repaired for another voyage. "Draw a chart and come back here," the czar had said, but he had also spoken of his interest in knowing about settlements in America. The captain decided that he could not return to St. Petersburg until he had made an effort to find the American shore, particularly since he was convinced that only a few days would be needed. "You can see land on the other side, when there is no fog," the natives had told him, and he had seen on the beaches evidence of another country close at hand—newly felled trees of a sort that did not grow in Kamchatka.

The *St. Gabriel* put to sea again the following June, and for three days Bering sailed her eastward in mist that never lifted, far enough to have reached any land visible from Kamchatka, he judged. On the fourth day he turned back without having sighted anything—he had missed the Aleutian chain.

Early in 1730, Bering was back in St. Petersburg, having retraced in just over six months the route over which his expedition had struggled for more than three years. His reception was anything but what he had expected. He had done the czar's bidding. He had the charts to prove it. But Peter had been dead for five years, and those who now controlled the government were as skeptical of the expedition's accomplishments as his officers had been on that late summer day when the *St. Gabriel* came about to sail back through Bering Strait. "You did not go far enough north. You did not sail west at all. You cannot be sure there is no land bridge," he was told. "Your charts are worthless."

Crestfallen, Bering nevertheless defended himself with spirit. Anna, Peter the Great's niece, who had come to the throne, took his part. After much wrangling with officials, she was responsible for the decision to go on with the plans for Siberian exploration and to retain Bering in command.

He was granted a new title, captain commander, and with the promotion came orders of a magnitude that must have staggered him: The entire coastline of Arctic Russia and Siberia from the White Sea to Kamchatka was to be charted. The interior was to be studied, its rivers charted, its people and resources reported. Lighthouses were to be erected, saltworks, foundries, and distilleries established. Schools were to be founded, and a mail service set up across those thousands of trackless miles from Moscow to Kamchatka. And after the beginnings of all that had been attended to by the captain commander, he was to dispatch an expedition from Kamchatka to locate the islands of Japan, and he was to build ships for his own exploration of the American coast as far south as the lands that Spain claimed.

The recently founded Imperial Academy of Sciences had played a large part in the planning, and Bering wondered if the scientists realized the extent of their ambitions. He assured them that the Great Northern Expedition would require an army of men, mountains of supplies, and he hesitated to guess how many years.

The men of science had pat answers to his objections: They were allowing four years for completion of the project. The country to be explored would supply manpower and materials. Distant towns and settlements were already getting the word about requirements. If the commander planned properly, he would begin sending out men now. Then, as he progressed across Siberia to supervise the organization of the various parties, he would find much of the groundwork done. They themselves were going out on the project in considerable force, the scientists informed him, by way of encouragement.

Although he was probably not elated with that piece of news, Bering must have been relieved to learn that the charting of the first section of coastline from the White Sea to the Ob' River was to be done under supervision of the Admiralty. All the rest of the grandiose scheme would be his responsibility.

When he left St. Petersburg in April, 1733, the captain commander was startled at the size of the caravan in which he rode. Wagon after wagon rolled out of the capital loaded with all manner of instruments, and there were other vehicles for the academicians, their secretaries, servants, and bodyguards. It was apparent, too, that the men of science were traveling with an eye to personal comfort. The baggage included tents, rugs, wines, and other niceties designed to make life in the wilderness endurable. If the rugged veteran felt any scorn for this retinue, he failed to mention it.

At Tobolsk, preparations began for the second charting expedition, which was to survey the coast from the Ob' to the Yenisey River in one season. After wintering, it was to go on northeast as far as possible. Bering ordered the construction of a seventy-foot sloop, the *Tobol,* and several small craft for provisions. The command was given to Lieutenant Ovtsin, and in May, 1734, the *Tobol* sailed on the first leg of the long journey to the Arctic.

Yakutsk next, thought Bering, and must have felt more like a general than a sea captain, as he prepared to put his army on the road again. He hoped earnestly that his advance forces would have the work to be done there well in hand before his arrival.

Four months later, his eyes widened as he crossed the plain toward the town. Yakutsk, the little walled village on the Lena River, which he had last seen on his way back to Russia in 1730, had spilled out of its walls in a wide, graceless sprawl of sod huts and half-finished log buildings. He soon learned that, aside from that, almost nothing had been accomplished. A furious winter of work lay ahead, if he was to keep to the schedule set for him.

Yakutsk was never a happy place during the years of the Great Northern Expedition. The demands upon the native tribes and the country's resources were so great that Bering was at times threatened with open rebellion. The scientists and officers bickered among themselves, and voiced their dissatisfactions in a rising volume of complaint against the captain commander. He was charged with everything from inefficiency to abducting a harem of native women, and St. Petersburg heard about it all via the postal service that he was maintaining with painful effort.

Despite such troubles, Bering managed to set up a foundry and a dockyard in which two sloops, the *Irkutsk* and the *Yakutsk,* and a number of provision boats were built during the winter of 1734–1735.

The plan called for the two sloops to sail in company down the Lena to the Arctic Ocean the following summer. The

Irkutsk, commanded by Lieutenant Lassinius, was then to make her way east along the coast and around East Cape to a rendezvous with Bering on the Kamchatkan coast, while the *Yakutsk,* under Lieutenant Prontschischev, sailed west around the most northerly point of the continent to a possible meeting with Lieutenant Ovtsin's expedition.

Bering saw the vessels push off in late July, and must have wondered a little at Prontschischev's wisdom when he saw the lieutenant's young bride waving from the *Yakutsk.* He noted that they were all as gay as though they were off on a summer's outing. Perhaps he envied them, as he turned away. He was a man of the sea and longed to be on his way to Okhotsk, where his own part in the great undertaking would begin. However, there were vast amounts of work still to be done in Yakutsk—barges to be built, posts and schools to be established, and the unending demands of the academicians to be met.

Bering was still in Yakutsk in the spring of 1736, when the first report on the two Lena expeditions came in: The Lassinius party was in serious trouble not far from the river mouth. Scurvy had attacked the *Irkutsk*'s crew during the winter. Lassinius was dead—most of the rest, too, said the one man who had found the strength to fight his way back up the river. There would be no living men at all if help did not come to them soon.

Bering sent off a relief force in haste, and with the rescuers went Lieutenant Dmitri Laptev to take over the sloop and her mission. A few weeks later, Bering saw the survivors of Lassinius' expedition, less than a dozen emaciated specters. Grimly he hoped that Prontschischev was faring better and wondered how long it would be before he learned the outcome of the *Yakutsk–Irkutsk* voyages.

As it happened, both vessels returned the following summer, but Bering heard only the report that Dmitri Laptev had to make. "It is impossible to sail along the coast and into the Pacific," the lieutenant asserted. "I went east until I came to two points where ice lies fast to the shore. The natives say that it never breaks up."

"Go to St. Petersburg with your report," Bering told him. "They must decide what they want to do up there now."

The two points Laptev mentioned were probably Syvatoi Nos and another just south of it, off Dmitri Laptev Strait. However, there must have been a misunderstanding between Laptev and the natives—no stretch along this coast has ice that prevents passage.

The captain commander had been having a bad time that summer, four years after his departure from the capital. Every post had brought official complaints about his lack of progress and the mounting costs of the expedition. The last straw had been a notice that his salary was being cut until he had achieved some of the objectives outlined for him.

When Laptev arrived with his report of failure, Bering was preparing to leave Yakutsk in order to make ready in Okhotsk for the Japanese voyage and his own to America. For all he knew, his time might be short. St. Petersburg might well decide to relieve him of command before he could sail.

By the time the *Yakutsk* reached the dockyard, the captain commander had taken the trail east, and so missed the tale of defeat and tragedy that Prontschischev's mate, Chelyuskin, related. The lieutenant was dead, buried on the shore of the Arctic, he said, and his young wife beside him. "They died last September while the ship was in the ice, he first and she two days later, and we carried them ashore when we could and said prayers over the graves. After that winter came—it was terrible, not like the winter before when we were all snug and happy in huts that we found on the bank of the Olonek River. We were so sure that first winter that when summer came we could sail on without any trouble.''

Chelyuskin shook his head when he was asked how far north the *Yakutsk* had sailed. He did not know exactly. "Above seventy degrees, I guess, and I can tell you that no one is going to sail any farther.''

Chelyuskin followed Laptev to St. Petersburg to make his report, and both men were sent back to Siberia to complete the work of their expeditions. Both were successful, although they were obliged to abandon the sea for land journeys, and years were required to achieve their goals. Laptev made his way overland from the Kolyma River on the Arctic to the Anadyr on the Pacific, and Chelyuskin sledged to Cape Chelyuskin, the northernmost point of the Old World.

With the completion of these journeys, the work of the Great Northern Expedition on the northeast sea route was ended. Thanks to the efforts of Vitus Bering, who never sailed any part of it and had only glimpsed the eastern extremity, its extent and general course had been charted.

Bering never knew this. While Chelyuskin and Laptev were toiling to finish their tasks, he was at sea on his last voyage of discovery. In 1741, he succeeded at last in reaching the shores of North America, but he was not among the handful of survivors of his ship, the *St. Peter,* who finally came back to Siberia. He had died and lay buried on an island in the Bering Sea.

Unhappily, the closing of the books on the Great Northern Expedition found the government in St. Petersburg expressing nothing but skepticism and ingratitude. Discoveries made at a fearful cost in lives were doubted. The separation of America and Asia had not been determined, it was said.

Spanberg, whom Bering had sent on a successful voyage to Japan, had not been there at all. Chelyuskin had not made his way afoot to the cape he had described.

Years were to pass before men of other nations came to the North Pacific and the Arctic to verify the findings of Captain Commander Bering's Great Northern Expedition.

8 *To Cape Chelyuskin and Beyond*

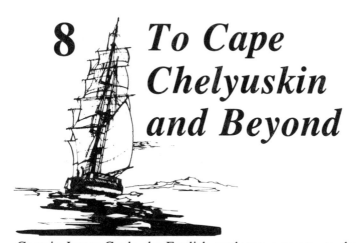

Captain James Cook, the English explorer, was among the first to realize that Russia owed Vitus Bering an immense debt of gratitude. When he reached Alaska in 1778, he found Russians engaged in a flourishing native commerce that had surely resulted from Bering's discovery. Late in the summer of that year, he took his ships *Resolution* and *Discovery* through the strait between Asia and America, and probing both east and west in search of an Arctic passage to the Atlantic, he proved beyond a doubt that what Bering had claimed fifty years earlier was true. The continents were not linked.

Captain Cook brought his ships back through Bering Strait, convinced that an interocean transit through northern waters was impossible. He deserves no more criticism for his failure to make the passage than do any of the early explorers. The means of achieving it did not yet exist. Almost a hundred years were to pass before it became clear that a steam-driven vessel might succeed where sailing ships had had no chance at all.

An Austrian expedition headed by Karl Weyprecht and

Julius von Payer was the first to try steam power on the Northeast Passage. Without mentioning any interest in sailing to the Pacific, they left Tromsø, Norway, in mid-July, 1872, in the little *Tegetthoff,* which boasted a small auxiliary steam engine. According to their stories, the expedition was setting forth merely to gain glory for the Austrian flag in the field of Arctic exploration. Later they admitted that a northeast transit had been the prime goal.

As it happened, 1872 was a bad ice year, and the *Tegetthoff* was beset in the pack ice off Novaya Zemlya five weeks after sailing. She never escaped. The explorers drifted north for more than a year until they sighted an unknown land which they claimed for the Austrian emperor, naming it Franz Josef Land. In the spring of 1874, they left their hopelessly entrapped ship, hauling sledges and boats, and after a wretched trip, reached safety on the coast of Norway three and a half months later.

No one learned of the Austrian expedition's return from their ordeal with greater interest than Professor Adolf Erik Nordenskjöld. Swedish geologist and Arctic explorer. He was sure that the explorers had intended to try for a passage to the Pacific, a thing that he himself had had notions of doing for some time. He wondered about the *Tegetthoff* expedition. If indeed he was right about their intentions, had they not headed too far north? Nordenskjöld had made himself an authority on the history of the search for a northeast passage, and he had queried hundreds of experienced walrus hunters and fishermen on ice conditions along those parts of the route which they frequented. His research had convinced him that an open-water passage close to the continental shore would be found during the late weeks of summer.

Payer had claimed that a northeast passage would be without commercial value, but Nordenskjöld was sure that at least one part would be of immediate worth. Certainly, a practical water route from the Atlantic to the mouth of Yenisey River would have immense advantages for European industrialists who had invested heavily in Siberian mining and timber properties. However, those to whom the Swedish explorer broached the subject of an exploratory voyage seldom failed to mention that attempts to cross the Kara Sea had never been successful.

"The way is open. Let me prove that I know what I am talking about," Nordenskjöld urged.

In 1875, he was given a chance with the little hunting sloop *Proeven* and sailed to the Yenisey without trouble. The following year he duplicated the feat with the steamer *Ymer,* which bore the first cargo of goods from Europe to Siberia.

Armed with this success, Nordenskjöld was ready to seek backing for a Northeast Passage expedition. Two merchants, one a Swede, the other Russian, who had been largely responsible for making the Yenisey voyages possible, were willing to contribute generously, but the costs of an undertaking of that scope would be beyond private means. The support of the Swedish government must be obtained.

At a lively session in the winter of 1877, Nordenskjöld summoned all his eloquence to the task of persuading King Oscar and his councillors to back his effort to find that northern passage between the oceans which had "never yet been ploughed by the keel of any vessel, and never yet seen the funnel of a steamer."

He stood before them, a man in his midforties, of medium height with a well-set-up figure and an intellectual's face deeply etched with lines of humor. Beneath a broad expanse of forehead his wide-set gray eyes were alight as he appealed to their national pride, noting the prestige that must result from exploration of this "immense expanse of ocean which stretches . . . from the mouth of the Yenisey past Cape Chelyuskin.''

"But in the time of Russia's Great Northern Expedition Chelyuskin stood on the cape and is said to have claimed that no vessel would ever pass it,'' someone objected.

"Chelyuskin was there in May,'' Nordenskjöld retorted. "At that time of year, it could not possibly have been ice-free.''

Another of the company remarked on the limited success of the vessels which Bering had dispatched from the Ob' and Lena Rivers to seek out the northernmost point of Asia. Nordenskjöld replied with a description of the vessels which had been sent down the rivers: "They were of flimsy con-struction, bound with willows, and caulked with a clay and moss mixture. The explorers never ventured very far from the coast. Indeed, they did not dare. . . . A properly equipped steam vessel would encounter no such difficulties as they had.''

Questioners would not leave the subject. They obliged the explorer to admit that almost nothing was known of the coastlines immediately east and west of Cape Chelyuskin.

He waved aside the objection as of no consequence. He was sure that surging river currents must drive the ice away from shore there in late summer, just as they did on the portions of the route with which he was personally familiar.

He expected no difficulty on the passage to the Yenisey mouth, said Nordenskjöld. After all, he had done it twice before. He would sail from there for Cape Chelyuskin about the middle of August, and if all went well, would be in Bering Strait a month later. Then he could complete a cir-cumnavigation of the Eurasian continent by way of Japan, the Indian Ocean, the Suez Canal and the Mediterranean Sea.

Knowing that he might have stunned his audience with the

magnitude of what he was proposing, he paused a moment, then added a sly remark about the recently opened Suez Canal. It was, he said, ''a splendid work which reminds us that what today is declared by experts to be impossible, is often carried into execution tomorrow.''

The explorer was so confident that doubt among his listeners gave way to enthusiasm. The king declared himself willing to contribute funds as a private individual. He would urge the government to grant support. The original backers promised heavier investments than they had planned. Professor Nordenskjöld might start planning his expedition at once.

Nordenskjöld lost no time in purchasing the *Vega,* a stoutly built little bark-rigged steam whaler. She had scarcely been signed over to him when he learned that his expedition was to be a far more ambitious undertaking than he had dreamed it might be. His backers, deciding not to neglect a possible profit from the venture, were going to place three other vessels under his command for parts of the passage.

According to the new plans, the steamer *Fraser* and the sailing ship *Express,* carrying coal reserves for the *Vega,* were to be escorted across the Kara Sea to the Yenisey, where they would load cargo for a return to Europe. The *Lena,* a small steamer designed to trade on the Lena River, would leave the *Vega* only after the most difficult part of the voyage, doubling Cape Chelyuskin, had been accomplished. Nordenskjöld reflected that not since the time of the sixteenth-century English and Dutch expeditions had such a flotilla prepared to sail off to the northeast.

The selection of the right men for the expedition was of critical importance. Viking-blond, bearded Louis Palander, Swedish naval officer and experienced Arctic voyager, was named the *Vega*'s captain. Other officers were a Swede, a

Dane, and an Italian, all from their national navies. Five civilians were selected to carry on the expedition's scientific studies, and the best of the volunteers from the Swedish Navy signed on as crew. All told, thirty men—officers, scientists, and crew—had found quarters aboard the little whaler when she sailed from Sweden July 4, 1878.

During the early part of the voyage Nordenskjöld followed the plan outlined in Stockholm almost exactly. Less than a month after her departure, the *Vega* led the expedition through the pass from the Barents Sea to the Kara Sea. However, fog and areas of broken ice slowed the crossing to Dickson Island at the Yenisey mouth, and upon arrival there, several days were required for transshipping coal reserves from the *Fraser* and the *Express*. Nordenskjöld saw no reason for concern about the loss of time. He was still ahead of his schedule.

On August 9, the *Fraser* and the *Express* weighed anchor for the trading voyage upriver, while the *Lena* and the *Vega* prepared to sail off into the unknown. Now the adventure was truly to begin.

From this point on, the explorer had no charts to guide him except those made at the time of the Great Northern Expedition, and he quickly found them to be less than trustworthy. During the first two days' sailing from Dickson Island, mist revealed one small island after another where none should have been. The two ships moved with extreme caution, sharp lookouts aloft, the leads going constantly in the bows. Twice, when the prevailing mist became very dense, they were obliged to lie to for a period.

The water had been clear of ice, as Nordenskjöld had foreseen, but he soon had enough of courting disaster on a blundering course through an uncharted island chain. They would draw away from the land a little, he decided. The new course brought them into an area of drifting ice, but the pieces were small and posed no hazard except when the fog closed down. Occasionally the ships tied alongside floes to wait out such a period.

The last of the halts on the northeasterly course came when Nordenskjöld was sure that they must be nearing Cape Chelyuskin. At that point, the ships were forced to swing at anchor in a small bay for four days. He was edgy about the delay and on the fifth day, when he persuaded himself that the fog was thinning a little, he gave the order to go on. The two vessels crept out of the bay, steaming slowly, breaking the Arctic stillness with repeated throaty whistle blasts when they lost sight of each other.

The explorer scarcely slept during the next day and a half. His excitement grew by the hour, for he knew that he was now very close to the cape. From time to time, he caught glimpses of a high rocky coast, and once he made out a sheet of ice lying fast to the land. He wondered if the cape itself would be so surrounded by ice that no landing would be possible.

Early in the afternoon of the second day's steaming from the bay, visibility improved, and soon a low dark land mass, completely ice-free, began to rise on the northeastern horizon. For a while Nordenskjöld watched it grow without speaking. Then, no longer in doubt about what he was seeing, he shouted his exultation: "That is Cape Chelyuskin. We have made it!"

Late that day, August 19, 1878, the *Vega* and the *Lena* entered the little bay that indents the northernmost promontory of the Old World. A single polar bear, pacing back and forth on the beach, eyed them curiously, as the anchors went down. Flags streamed to the tops, while the ships' guns were readied to salute the occasion. At the first blast, the bear took to his heels. Nordenskjöld laughed at the hunters who had thought to bag a fine trophy at the expedition's farthest-north anchorage.

The land about the cape was as monotonous and unlovely as any of the explorers had ever seen. A gray desolation of fragmented rock, marked here and there by fields of snow, it rose gradually to low hills farther inland. In a few places there was a stunted, scanty ground cover.

No animals were seen after the bear's disappearance, but the sky was full of migrating birds—geese, sandpipers, gulls, and kittiwakes. They have perhaps come from some unknown land beyond here, thought Nordenskjöld. He had no idea that his ships had passed through a narrow strait to reach the cape, and that another shore lay close at hand to the north. Severnaya Zemlya was to remain undiscovered for another thirty-two years.

There was nothing in that grim landscape to encourage prolonged stay, so the scientists made haste with their scheduled tasks. After only thirty-six hours the expedition left the cape astern, steaming east into the Laptev Sea.

Captain Palander probably raised his eyebrows a little at this change in the original plan to follow the coast in a southeasterly direction.

"If we are able to hold this course, we may fall in with new lands on the way," Nordenskjöld explained. "And it will be a much shorter passage to Bering Strait, of course."

The comparative ease with which the expedition had gained Cape Chelyuskin may have made him a little overconfident. He was soon to learn that the Laptev Sea bore little resemblance to the Kara Sea.

9 The Vega Sails On Around

Two days after the *Vega* and the *Lena* had left Cape Chelyus-kin, they were surrounded by drift ice in fog "so thick you could cut it with a knife." Nordenskjöld knew that he had made what might be a fatal error in deciding to sail east from the cape.

"We've got to get out of this and make for the open water along the coast, or we'll end up as the *Tegetthoff* did," he told Palander.

The captain tried east, south, and west for an exit from the ice labyrinth and was balked at every turn, until he found at last the small opening through which he had sailed more than thirty hours earlier. Two hours later land was in sight, and presently, the two ships were moving southeast along the coast without trouble.

Nordenskjöld remembered that Prontschischev had pioneered this track almost a century and a half before. Perhaps he had been very close to Cape Chelyuskin before illness and the fear of being beset had turned him back. His own observations at the cape seemed to bear out the accuracy of Mate Chelyuskin's guess that the *Yakutsk* had been above

70 degrees north. (The *Yakutsk* had, in fact, been a little more than three hundred miles from Cape Chelyuskin.)

On the night of August 27, the *Vega* and the *Lena* parted off the great delta of the Lena River, the *Vega* firing rockets in farewell salute to the little consort ship, as she turned into the river which was to be her permanent home.

When the *Vega* sailed alone the next morning, she left the coast on a northeasterly heading. Things had gone so well for the past few days that Nordenskjöld thought he could spare the time for a brief excursion into the New Siberian Islands, reputed to be a treasure house of ivory from the tusks of mammoths, the huge, elephantlike beasts that had roamed the Arctic regions in prehistoric times.

The ivory-hunting expedition was fruitless. The waters about the islands proved to be so shallow that often the ship's leadline measured little more than two fathoms. The anchorages were open and extremely hazardous. Nordenskjöld dared not risk putting a boat ashore. On the last day of August, he reluctantly turned his back on the New Siberian Islands and again set out for Bering Strait.

For the next forty-eight hours the *Vega* made splendid progress in fine weather, but as she approached the Bear Islands, it became apparent that the Arctic winter was setting in. Snow fell, powdering the decks, and drift ice began to appear. During the nights a thick crust formed between the old pieces, often bringing the ship to a halt. On September 4, Nordenskjöld was again obliged to order a turn toward the coast.

Captain Palander had no easy time that day. The ship's wheel spun without pause, as he called the orders for a twisting course through the drift, and a few good solid shocks were felt throughout the ship when he decided to ram chunks aside. He must have breathed a sigh of relief when, close to nightfall, he sighted a broad belt of open water extending along the shore.

Nordenskjöld's journal entries for the next two days note "satisfactory progress," but they reflect the fatigue that everyone aboard the ship was feeling. "Even the most zealous Polar traveler may tire at last of mere ice, shallow water, and fog," he wrote, and commented on the monotony of the voyage at that point.

Shortly the *Vega*'s pace slowed to a crawl, and there was no further mention of "satisfactory progress." Days passed when she did not move at all. Nordenskjöld's anxiety grew, and he began to ask himself questions: Was this one of those bad ice years that spelled defeat for the best of Arctic voyagers? Or was he too late—perhaps only a few days too late? If that was so, could he have made better time?

His answer to the last question was honest. The operation at Dickson Island could have been hurried. He could possibly have left that bay west of Cape Chelyuskin a bit earlier. Above all, he regretted the time spent in the New Siberian Islands.

The *Vega* came to a final halt at the end of September, not more than a day's steaming from Bering Strait, under good conditions. For a short time Nordenskjöld hoped that a south wind might break up the pack that imprisoned her, but when it failed to rise, he was obliged to accept the fact that his ship was going to remain where she was for a long time.

Everyone aboard ship shared his disappointment, but there was no sense of crisis. This would be no such bitter wintering as had been endured by many of the early explorers. The planning had been too good for that. There was an ample supply of warm clothing in the hold and enough food to last two years. Moreover, the cook would have nothing to complain about in the variety supplied for his galley. Nordenskjöld had made his selections with care, knowing that scurvy was not a problem in a ship that provided a proper diet for its men.

He was a bit uneasy about the *Vega*'s position. She lay in the shallow water of the open road, a little less than a mile from shore, and might well be crushed by the violence of winter storms. He was certain that she could be abandoned safely, but he was to spend many a sleepless night worrying about the cracks and groans with which she resisted ice pressure.

News of the steamer lying offshore had raced along the coast, and as the days passed, Chukchis appeared on the beach in increasing numbers. From time to time dog sledges drove a short way onto the ice. Clearly, the object was to get out to the ship, but at first the surface crust was too thin to venture upon. After the first week of October, however, there was a steady traffic between ship and shore. Sledges, with harnessed dogs, curled nose to tail in the snow, lay alongside the *Vega* all day, while their owners chattered, begged, and bartered on deck. And the ship's people began to make regular trips over the ice to investigate the terrain and further their acquaintance with the inhabitants of several nearby villages.

Nordenskjöld's interest in conditions ashore was far from casual. He knew that good morale during the long night of winter would depend upon a program of useful activity for all hands. Moreover, the work he had in mind could not fail to make a valuable contribution to knowledge of the Arctic.

A short time after the *Vega* was beset, he called a meeting of the officers and professional men, and ticked off his plans for the winter in the order of their importance: ''We'll build an ice block observatory for scientific instruments back of the beach. It will require a twenty-four-hour watch. The people of this area, their customs, clothing, and habitations should be the subject of study. The opportunity to explore as much of the Chukchi Peninsula as possible must not be neglected. Those of you who wish may plan journeys of some extent, if dog sledges and guides are available, and I have no doubt they will be.''

He smiled, thinking of the eagerness with which the Chukchis were trading for the treats to be had from the ship. ''Finally, this should be a time for self-improvement among the members of the crew. We have a library of a thousand books aboard. They should be encouraged to read, and you gentlemen ought to be planning lectures on your special fields.''

A routine was established that allowed little time for boredom. Each man had a watch to keep and tasks to perform, but there was time for recreation, too—ice skating and visiting ashore during the day, while music, games, and reading passed the hours after supper.

Little excuse was needed for a party. All anniversaries and birthdays were celebrated with songs, toasts, and something special on the supper menu. Christmas was, of course, the most festive occasion of all. A tree, fashioned of a piece of driftwood and brush-willow branches, was decorated with flags and small candles, and dozens of gift parcels (provided by Nordenskjöld before departure, in the remote chance of a wintering) were brought up from the hold. Men paired off and danced many a clumsy, merry polka around the tree on Christmas night, while the holiday punch bowl was emptied with toasts to king, country, and the families at home in Sweden.

On the whole, the *Vega*'s people were content enough while they waited for spring to free them. Without fear of hunger or the Arctic cold, they were only uneasy when ice, buckled and tossed by the waves of a distant storm, squeezed the ship's hull to noisy protest.

After the first such experience in mid-December, Nordenskjöld decided that wisdom called for a supply cache on the beach.

The cache was never needed, and secured only by a sailcloth covering, it remained undisturbed, even though people who were always hungry passed it every day. The Chukchis had had a poor hunting and fishing season, and were truly facing starvation. Unfortunate as it may have been for the expedition, the wintering proved to be a blessing for them. Had it not been for the food supplied by the ship, many in the villages might not have survived until spring.

February brought a period of mild weather, and in the weeks following there were occasional flurries of excitement aboard the *Vega* when evidences of an early breakup were observed. Nordenskjöld canceled plans for extended journeys ashore, so that the ship could move without delay in that event. However, each time a crack in the ice seemed to promise an open lead, it closed up promptly, and as winter drifted into spring, the weather grew more severe than ever. April was the coldest month of all.

The appearance of migratory fowl was the first indication of spring's reluctant approach. Nordenskjöld's journal noted a flight of snow buntings on April 23. They were followed by geese, ducks, gulls, waders, and songbirds. In May, huge flocks of Arctic warblers settled in the *Vega*'s rigging to rest after the long flight from their wintering rounds in southern Asia.

Patches of bare earth began to spot the snowy landscape, and cracks appeared more frequently in the ice field. Early in June, the Chukchis said that the first whalers had reached a point only a little distance to the east. But still the *Vega* lay fast in her prison.

"Will she never come free?" men began to ask when the calendars were turned to July. Nordenskjöld himself, when he sat down to the midday meal on July 18, saw no possibility of release before the end of the month. He was talking of going ashore that afternoon when he broke off in midsentence. There had been a slight jar.

"She's moving!" Palander cried and rushed on deck. The ice had broken away from the sides of the ship and was in motion. "Fire the boilers!" he shouted.

At 3:30 P.M. steam was up, and the *Vega* moved slowly away from the place of her long imprisonment. All her flags flew in salute to a forlorn group of Chukchis who had gathered on high ground to watch their friends of the winter disappear forever. Nordenskjöld wondered if they were crying, as they had often said they would on this day. He thought it not unlikely, and confessed to a moment of sadness at the separation.

Fog shrouded most of the coastline for the rest of the *Vega*'s voyage through Arctic waters, but Nordenskjöld was able to make out the dark heights of East Cape (Cape Dezhnev) on the morning of the twentieth. By 11 A.M. the ship was in the middle of Bering Strait.

It was a solemn moment for the explorers. "Three hundred and twenty-six years after it was first attempted, a ship has at last traversed the Northeast Passage," said their leader. Again he ordered flags flown and a salute fired, and spoke of his pride in seeing Sweden's blue-and-yellow ensign rise to the masthead, "where the Old and the New Worlds reach hands to each other."

There was no haste to complete the voyage. Nordenskjöld

took time to visit points of interest on both the American and Asian shores of the Bering Sea. Upon reaching Japan, the *Vega* required a month for repairs, and the crew spent another month in junketing about there and in China, in order to wait out the monsoon season in the Indian Ocean.

There were halts at Singapore, Aden, and Port Said. Festive receptions prepared at Naples and Lisbon for the Swedish explorers could not be passed by. England demanded a chance to honor the little vessel and her men. So, too, did France and Denmark. During the final weeks of the voyage, Nordenskjöld, exhausted and longing for home, found little to write in his journal but long lists of official receptions, banquets, and honors conferred.

Fourteen months after the *Vega*'s triumphant salute to Bering Strait, she at last entered the harbor of Stockholm, her place in history ensured as the first vessel to sail completely around the Eurasian continent.

10 The Polar Ship U.S.S. Jeannette

The *Vega*'s failure to appear anywhere in the North Pacific during the late months of 1878 was a source of worldwide concern. The *Lena*, upon reaching a river port in Siberia, had sent the last information on her whereabouts. The fact that she had not emerged from the Arctic Ocean could only mean an enforced wintering or disaster somewhere along the Siberian coast between the Lena River mouth and Bering Strait. Sweden made haste to send the steamer *A. E. Nordenskjöld* to the relief of her missing explorers.

In the spring of 1879, while the *Nordenskjöld* was still on her way to the Pacific by the southern route, the United States Secretary of Navy realized that he was in a position to offer the Swedes aid of real consequence in the search for the *Vega*.

He had before him the proposal of James Gordon Bennett, publisher of the *New York Herald,* to make a naval project of an expedition to reach the North Pole via Bering Strait. Bennett promised to purchase the vessel, and pay all of the other costs, if she was commissioned as a ship of the United States Navy and manned by naval personnel. The Secretary decided that he would approve the plan, on condition that the expedition's first business be a search for the Swedish explorers.

United States Navy Lieutenant George Washington De Long, Bennett's expedition leader, was not happy about the edict that he must squander good summer weather in poking about for the *Vega* before he could turn north. The delay might well rule out his polar venture until the next year. Nevertheless, the orders could not be ignored, so during the last week of August, 1879, the *New York Herald* ship, U.S.S. *Jeannette,* put into St. Lawrence Bay on the Siberian coast south of Bering Strait to begin the search.

There was no sign of life on the bay until a couple of skin boats loaded with Chukchis pushed away from shore and made for the *Jeannette*. The natives swarmed aboard uninvited, chattering madly in a language that boasted at least a smattering of English. De Long put his questions in the simplest words he could find: Had they seen a ship something like this one lately? Vigorous nods of assent. Her name? Blank looks. Her captain's name? Again no response. The name of any man in that ship? "Horpish!" one of the visitors brought forth in triumph, and told a barely comprehensible story of a winter journey across the Chukchi Peninsula to the Arctic Ocean, where he had seen the same ship frozen in.

"It must have been the *Vega,*" said De Long, "but what about this Horpish? There was no one of that name on the ship." He ran his finger down the *Vega*'s crew list again, stopped at Nordquist. "You know Nordquist?" he asked. "Horpish!" said the Chukchi, head bobbing in happy agreement.

This seemed to be evidence enough that the *Vega* had completed her passage not too long before, but it was not conclusive. De Long sailed at once for the place described by the native. There, the search ended with the discovery of Swedish coins, uniform buttons, and other items which could only have come from Nordenskjöld's vessel. The *Jeannette*'s bow was promptly turned toward the Pole.

George Washington De Long's previous Arctic experience had amounted to no more than taking part in a naval search for a vessel missing in Baffin Bay in 1873, but that brief taste of adventuring in northern waters had been enough to inspire dreams of heading an expedition to reach the North Pole. In the years since, he had made himself familiar with the history of polar voyaging, and he had pondered every scrap of theory put forth by Arctic authorities to account for the lack of success thus far in gaining the farthest-north point of the globe.

All of the earlier expeditions had sought a route north from the Atlantic. Some of the experts had begun to speculate that the Pacific was the proper avenue of approach. They cited the belief that the Japanese Current split in the North Pacific, with one branch turning south to warm the northwest coast of the United States, and the other flowing through Bering Strait into the Arctic Ocean and on past Wrangel Island. Indeed, the current's warmer water might create an ice-free passage all the way to the Pole, said the armchair explorers. And, in case this was not so, sledging parties should be able to reach the Pole from a Wrangel Island base.

This was the plan that De Long had sold to James Gordon Bennett for his polar expedition, but he had not been long in the Arctic before he began to suspect that the Japanese-Current theory was in error. As he sailed north, he found no open water channel—only more and more ice. The ice pilot, who had been engaged at an Alaskan port, was not surprised. There might be a way farther east, he thought, but he had known all along that there would not be one in the direction of

Wrangel Island. Six days after leaving the Siberian coast, De Long's ship was trapped in the pack.

Originally a British naval vessel, the *Jeannette* had been purchased in Europe and turned over to the United States Navy for conversion. Shipwrights at Mare Island Navy Yard on San Francisco Bay, where the work was to be done, shook their heads over her. "Not suitable," they said, but they did the best they could. When they had finished, her sides were nineteen inches thick, and her bow solid wood for ten feet abaft the stem. Wrought iron sheathed the bow, and great beams had been bolted athwartship to strengthen her laterally. Surely she could resist any amount of pressure in the ice, thought De Long, and did not listen when the Mare Island people spoke of hull lines that could not be altered. They saw her sail with misgivings.

Nordenskjöld had known periods of anxiety when his ship, squeezed in the ice, had cried out in tortured protest, but he never experienced anything like the terror that reigned in the *Jeannette* for weeks on end during the winter of 1879–1880. Packed sledges stood on deck, ready for instant evacuation, and men slept in their clothing when the ruptured pack sent blocks of ice, some as big as two-story buildings, crashing over each other with an incredible uproar.

Time after time, it seemed that the ship must be overwhelmed by tons of moving ice, but each time the deadly avalanche passed her by, or unbelievably, stopped its movement just short of catastrophe.

The recurring periods of near panic did nothing to ease growing discord in the ship, and De Long greeted the summer of 1880 with relief. Quarreling over duties and privileges would stop once the ship had broken out of the ice and the northward voyage had been resumed, he told himself. But June and July brought no change in the pack. "The breakup will come in August,'' the captain said with confidence. However, it did not. At the end of the month, he ordered the ship prepared for another winter.

To add to his bitter disappointment, he knew that, after a full year, the *Jeannette* was a mere 150 miles north of where she had been beset. During the first months, when the pack had drifted northwest at two miles a day, he had hoped to break out reasonably close to the Pole. But the drift had reversed, sending the ship back along the way she had come.

Again, through months of terror-filled darkness, the *Jeannette* drifted northwest with the pack. Illness mounted in the ship, and squabbling reached the point of open warfare. Only De Long still talked of a successful completion of the mission. He was counting on enough time during the following summer to reach the Pole—after the ice had released the ship from its grip, of course.

He wasted no time on the notion that the *Jeannette* might be a permanent victim of the pack. In fact, as spring came on, he grew a bit nonchalant about its menace. The ship had survived so many threats of destruction that she seemed to lead a charmed life. Thus, he was ill prepared for the events of June 11, 1881, when the ice claimed her at last.

Ironically, that day began with a promise of freedom. A series of reports like heavy gunfire roused the ship early. The pack was breaking. Cracks flared away from the *Jeannette* in every direction. She tossed in her ice cradle as though on land rocked by an earthquake, lurching finally into a little space of open water that widened rapidly into a bay. In happy anticipation of a day that he had awaited for almost two years, De Long had the rudder reshipped and the rigging cleared for sailing.

His elation soon changed to horror. The bay was shrinking, the ice advancing. So thick that its lower edge lay deeper in the water than the *Jeannette*'s keel, it drew steadily nearer to her sides. The old ice pilot, whose counsel had been little needed, was gloomy. "Before tonight, she'll either be under this floe or on top of it," he told De Long.

The Arctic veteran was right. The ice closed in relentlessly and gripped the hull whose shape had worried the Mare Island shipwrights. The *Jeannette* began to give way to the pressure, shattering timber by timber, like a nut between the jaws of a nutcracker. By evening, she lay on her side on the floe, a mass of crushed wreckage that sank early the next day. Thirty-three men, huddled on the pack with their heaps of salvage, saw her go—the last link with home—and began a grim reckoning of their chances of ever getting back there.

They were five hundred miles north of the Lena Delta on the Siberian coast, and none of the castaways doubted that they would be miles of fearful travel, first across the chaotically tumbled pack, broken by leads of black water, and, after the southern edge of the ice had been reached, across the open Arctic to the mainland. It was useless to think of relief from any quarter. No one knew where they were.

De Long tried to hearten his company by pointing out on the chart of the Lena Delta numerous villages where help would be found, and spoke cheerfully of the prospects of reaching it. They had saved the sledges and the ship's three boats. They had warm clothing and provisions for sixty days, a quantity that would surely be sufficient for the journey. Their situation was far from hopeless, he asserted. They had only to put their shoulders to the task of sledging those boats over the pack, so that they could make the ocean crossing. He shrugged off a question about the extent of the pack. He had no idea how far it might reach.

So, the dreadful trek to Siberia over that nightmare icescape began. It ended in tragedy.

The *Jeannette* had been on the bottom for three months before her crew finally reached the open Arctic, ninety-six miles from the mainland coast. Clothing and boots had long since been reduced to tatters. The food, doled out over the weeks in smaller and smaller portions, was all but gone. Many of the men who set sail on the morning of September 12 were at the point of death from cold, exposure, and starvation. Still, their spirits were high as the sails filled for the crossing to Siberia. If the good east wind that was blowing held, another twenty-four hours would see them in a safe haven, fed and warm.

The early part of the day did nothing to diminish their feeling that the ordeal was almost at an end, but in the afternoon the wind freshened, backing into the north, so that the boats were constantly in danger of being swamped by following seas which built up by the hour. Evening found them plunging through a roaring gale.

The smallest of the boats, a twenty-footer in command of the *Jeannette*'s first officer, Lieutenant Charles Chipp, fell far behind, plainly in trouble. The men in the other two boats tried to check their speed, so that the flotilla would not be separated. Watching, as Chipp's boat struggled on, they saw a wave break over her stern. She broached, falling into the trough with useless sail and yard slamming wildly. For another moment she was visible, broadside on the crest of the next sea. Then she disappeared. In the turmoil of those mountainous waves, there could be no thought of rescuing the eight men who had sailed in her.

The two remaining boats, their courses dictated by the fury of the storm, lost sight of each other during the night. Eventually, the one in command of the *Jeannette*'s chief engineer, George Melville, reached a river mouth at the southeastern edge of the delta, while De Long's beached far to the northwest. Both parties discovered that they had landed in country as desolate and uninhabited as the moon. There were no welcoming natives, no smoke of village fires. The map-maker who had indicated settlements on the Lena Delta could not have known much about the Arctic.

Scarcely able to handle the oars, the eleven men in Melville's boat struggled upriver, searching in vain for signs of humanity or for game. They were very near the end of their endurance on the fourth day when they came upon three natives who gave them food and guided them to a native village, which was still a week's journey distant.

In the meantime, De Long had started south afoot with thirteen men, some of them still wrangling among themselves. Several were so weak they could scarcely keep their feet. There was no food of any sort to be found, and with the end of September, the Siberian winter set in in earnest. A man died, and De Long sent the two strongest of his companions ahead to get help. (He believed that they were about twelve miles from a village. Actually, it was more than seventy miles distant, and when the pair finally reached it, they could not make the natives understand their frantic pleas to send a rescue party back to the men they had left.)

While awaiting the help that never came, De Long's company continued at a snail's pace, trying at last to subsist on boiled willow leaves. In mid-October, they began to die. The last entry in the captain's journal, found with the bodies the following spring, revealed that on October 30 only he and the ship's doctor were still alive. They probably did not see another day.

The pitiful handful of survivors, ten of Melville's group and two of De Long's, reached home at last to tell their stories of the disaster, and the books were closed on the *New*

York Herald polar expedition. But not quite. The *Jeannette* was still to be heard from, and in a way that justifies telling her story with all the others about the Northeast Passage.

11 The Drift of the Fram

Three years after the *Jeannette* disappeared beneath the ice, an Eskimo living on the southwest coast of Greenland carried home a number of articles he had found frozen on a shorebound floe. By good fortune, these items, which could have had little value for him, fell into the hands of people who understood their significance. There were two pieces of paper: A list of provisions, signed by G. W. De Long, and a list of the *Jeannette*'s boats. There was a pair of oilskin pants marked ''Louis Noros,'' and the bill of a cap bearing the name ''F. C. Lindeman.'' Both men had sailed in the *Jeannette*. There could be no doubt that this was flotsam from the ill-fated ship.

But how had these things come to a shore thousands of miles from the place where the ship had gone down? it was asked. There was all sorts of speculation, but most Arctic authorities agreed that, in the course of three years, the floe bearing them had drifted clear across the Arctic Basin and into the Atlantic.

No one accepted this theory more enthusiastically than Fridtjof Nansen, a young Norwegian who dreamed of explor-

82

ing in the Arctic. The *Jeannette*'s relics had confirmed his notion that a westward current flowed somewhere between the North Pole and Franz Josef Land all the way from the Siberian Arctic to Greenland. In fact, he believed that the current must move very close to, if not over the Pole. If you wanted to reach that as yet unattained goal, as Nansen very much did, that was the route, he asserted.

One must work *with* and not *against* the forces of nature, he claimed, and working with nature meant allowing a vessel to be frozen in the pack ice north of Siberia, then simply waiting to be carried over the Pole.

There were all sorts of questions when the young man began to seek support for such an expedition. Would not his ship be sealed in the ice for a very long time? doubters asked. "Certainly," he replied, "perhaps for five years." But that would not matter, if a small crew of carefully selected men expected a long trip and were well provided for.

And what about the ship? Why might it not be crushed in the ice, as the *Jeannette* had been? Nansen answered that objection with a reference to the American ship's "preposterous hull," and described the vessel he had in mind: "I propose to have a ship built as small and as strong as possible . . . just big enough to contain supplies of coals and provisions for twelve men for five years. . . . The main point in this vessel is that it be built on such principles as to enable it to withstand the pressure of the ice. The sides must slope sufficiently to prevent the ice, when it presses together, from getting firm hold of the hull, as was the case with the *Jeannette*. . . . Instead of nipping the ship, the ice must raise it up out of the water. . . ."

He admitted with some humor that his stubby little vessel would be anything but comfortable underway, "but it would not be so bad a sea-boat as to be unable to get along, even though some sea-sick passengers might have to offer sacrifices to the gods of the sea."

He would sail the northeast route pioneered by the *Vega* as far as the New Siberian Islands, he said, and then turn north until the ship had found a cradle in the pack. After that, the motive power would be the current and the ship would become a barracks, until it emerged from the ice somewhere in the vicinity of Greenland.

Almost without exception, old Arctic hands condemned the idea as "sheer madness," but Nansen, heedless of such criticism, continued to press for financial backing.

Years passed, and he was Dr. Nansen, distinguished scientist and the veteran of a Greenland traverse, before his dream began to assume reality with the building of the *Fram* ("forward" in Norwegian), the stout little ship which he claimed would "be able to slip like an eel out of the embraces of the ice."

Early in the summer of 1893, Nansen, with twelve companions, sailed from Norway to begin his odyssey in the ice. From the beginning the *Fram* was everything, both good and bad, that he had expected her to be. He was among the first to fulfill his prophecy about her effect on passengers when she was at sea, but a few days later, he was delighted with her behavior in the drift ice encountered in the Barents Sea. "She twists and turns like a ball on a platter," he observed to his captain, Otto Sverdrup.

His pleasure in the ship's performance was tempered with dismay at finding ice so early in the voyage, however. At almost the same date in 1878, Nordenskjöld had seen none at all until he entered the Kara Sea. Knowing that if there was ice west of Novaya Zemlya, there would be much more east of the island chain, Nansen began to suspect that he would not be able to equal Nordenskjöld's time on the passage to

Cape Chelyuskin. In this suspicion he was proved right. The *Fram* logged thirty-seven days on the voyage from Novaya Zemlya to the cape, a distance that the *Vega* had covered in less than three weeks. There were times when the ice seemed determined to compel a first wintering in the Kara Sea.

On September 10, the *Fram* passed the northernmost point of the Old World, the third vessel in history to do so. She did not anchor in the little bay where the *Vega* and the *Lena* had lain, however. The season of open water was rapidly drawing to a close, and Nansen was anxious to get as far north of the New Siberian Islands as possible before committing his ship to the pack.

A week later, the *Fram* was on a northerly course, with one island of the group in sight to the east. Nansen was jubilant. Now at last everything was going according to plan. "Open sea; good wind from the west; good progress . . . Now the decisive moment approaches," he noted in his journal on September 18.

The *Fram* continued to forge on, and always the crow's-nest lookout had the same reply to hails from the deck: "Nothing but clear water!" The islands were left astern. The latitude in which the *Jeannette* had gone down was passed. "I have never had such a splendid sail," said Nansen and joked with Sverdrup about sailing all the way to the Pole.

He must really have given some thought to the possibility, for his journal entry of September 20 begins: "I have had a rough awakening from my dream." He had been below that morning, when a sudden clattering of yards over his head sent him rushing on deck. The *Fram* had reached the edge of the polar pack.

Sverdrup investigated an opening in the ice. It proved to be a bay. The *Fram*'s splendid sail had ended. Here she would remain to be frozen into the pack, Nansen decided, and

settled down to await confirmation of his long-cherished
theories about a northwest drift.

He could scarcely believe the results of the first observations. The *Fram* was going backward! He was in despair.
"We are drifting southwest," he wrote. "What can be the
reason of it? . . . I cannot account for any south-going
current here—there ought to be a north-going one."

An aimless movement, first in one direction, then another,
continued for the rest of the year, but early in 1894, the
situation improved a bit. The ship was at last a little north of
her first position in the pack, and a westerly trend was noted.
Still, Nansen complained of "a wretched state of affairs,"
and drew a melancholy conclusion: "At best, if things go on
as they are doing now, we shall be home in eight years."

Unhappy as he was with his progress, the explorer was
more than content with the way the ship had behaved. The
pressure against her sides, as the ice closed in, had not been
disturbing. She had lifted, just as he had said she would, with
ice thickening beneath her rounded bottom, until she rested
on a bed more than thirty feet through.

The *Fram,* thus imprisoned, was a tranquil ship, with one
exception. Nansen's journal reveals him to have been the
single malcontent in a snug, well-fed cheerful company. As
the slow months of the drift passed, and his suspicion grew
that the ship was not going to reach the vicinity of the Pole, he
often sat alone in his cabin, trying to write away his frustration and unhappiness. He told himself that his urgent desire
to stand at the very top of the world arose from personal
vanity, and was child's play in comparison with what he had
set out to prove. Still, he could not beat down his obsession.

He began to toy at length with an idea that had always been
in the back of his mind: Whenever the pack had carried the

ship to a position north of Franz Josef Land, he would leave it to make a sledge journey to the Pole.

The fall of 1894 found the *Fram* halfway between the New Siberian Islands and Franz Josef Land. The westerly drift was established, and its pace was accelerating. Nansen started to plan in earnest for a northward dash. He and one companion would leave early in the following year. The distance to be covered would then be less than five hundred miles. On the basis of his experience in Greenland, he thought the journey could be made in fifty days. The two men would not try to rejoin the *Fram*—the return would be made by way of Franz Josef Land and Spitsbergen. Before the winter of 1895 had set in, they should be safely home in Norway.

He broached the subject to Sverdrup. The captain thought it an excellent idea, urged that it be carried out. Would the leader of the expedition be criticized for leaving it? Nansen asked. "Nonsense," said Sverdrup. Eleven men could handle the ship very well, and he did not doubt his own competence to bring her through safely.

Heartened by the captain's encouragement, Nansen devoted the winter weeks to preparing for his expedition. As soon as news of the plan was known in the ship, everyone wanted to go with him. He selected Hjalmar Johansen at last, and together they worked out their lists of equipment and supplies.

On March 14, Nansen and his companion left the *Fram* with twenty-eight dogs and three carefully packed sledges, two of which bore lightweight kayaks for the over-water portions of the return journey. They were positive that their thorough preparations would ensure success.

For the first ten days all went well, as they skimmed

northward over level plains of ice. Nansen thought it seemed that they must extend clear to the Pole. "If this goes on, the whole thing will be done in no time," he told Johansen.

He spoke too soon. Only a day or so later, the plains gave way to pressure ridges of tumbled ice. Progress was slowed to a few exhausting miles a day. As early as April 3, the two men shared doubts about the wisdom of continuing.

Five days later, Nansen knew that he was beaten. With half of the estimated time for reaching the Pole used up, they had covered only a fraction of the distance, and there was nothing in sight but "a veritable chaos of ice-blocks, stretching as far as the horizon." Taking what comfort they could in having advanced farther north than any man before them, they must turn and make for Franz Josef Land, if they were to get out of the Arctic before summer's end.

The southward journey proved to be even more difficult than the advance, for when they again reached the plains, they found them broken by leads of open water that must be detoured or ferried over on ice blocks. And it was a far longer trek to land than Nansen had counted on. His observations and the maps made by Weyprecht and Payer twenty years before had led him to believe that land would be in sight early in May, but as the month wore on, it failed to appear in the southern distance. He admitted, at last, that he was thoroughly puzzled about their position.

Weeks passed while they continued to crawl forward under fearful conditions. Their food was almost gone, and the weakening dogs had to be destroyed, one by one.

Nansen did not permit himself to despair. "If the worst should come to the worst, and we have to make up our minds to winter up here, we can face that," he announced with stout courage. He managed to kill a seal and then a polar bear. They would not starve as long as they were able to hunt.

Finally, late in July, after more than four months on the ice, they glimpsed a cliff on the far horizon. Thirteen days later, they reached the northern shore of Franz Josef Land, and saw beyond it nothing but open water. They were jubilant. The kayaks would carry them swiftly south through the archipelago, then across to Spitsbergen, where they would find many fishing vessels. The dream of being home by fall would yet be realized. They destroyed the two dogs that had survived the march and launched their little craft with soaring spirits.

The picture changed in the space of a single August night. While the explorers slept on the shore of an island, the ice moved in, trapping them. After that, wintering was inevitable. They cobbled up a stone hut, roofed it with the skins of walruses that appeared offshore, and settled down to wait for the spring of 1896. Only the fact that polar bears roamed within range of their guns saved them from starvation during the dreadful months that followed.

May 19 saw them underway once more, tattered, incredibly grimy, but quite fit. They continued to push on south, and in mid-June were still counting on many weeks of lonely wandering when the voyage came to an abrupt end. Nansen had been exploring the terrain around their night's campsite when he heard the distant barking of dogs, a sound so unexpected that he could scarcely credit his ears. He streaked off toward the source to find that, by happy chance, he had stumbled upon the base camp of an English expedition to Franz Josef Land.

In his joy at seeing salvation at hand, he failed to identify himself to the first man he confronted, and the Englishman saw only a wild man clearly in need of succor. The creature was "clad in dirty rags, black with oil and soot, with long uncombed hair and shaggy beard. . . ."

He could not imagine where this savage had come from, but finally the light dawned. "Aren't you Nansen?" he stuttered, and upon hearing the reply, "By Jove! I am glad to see you!" The fate of the Norwegian expedition, unreported for almost three years, had been a matter of wide speculation. Now at last part of the mystery had been solved.

The rest of the *Fram* story was to unfold a couple of months later. Sverdrup pushed her out of the ice near Spitsbergen on the very day that Nansen and Johansen reached Norway aboard the English expedition's supply vessel.

Nansen was beside himself with joy. All hands were safe and well. The ship had proved herself in accomplishing the task he had set for her—the drift from the Siberian Arctic into the Atlantic. He had little to say about his disappointment in not reaching the North Pole, claiming instead the honor of having been, with Johansen, the first to go beyond 86 degrees north latitude.

In his satisfaction with such a pair of distinguished "firsts" in the history of Arctic exploration, he found no time for mention of an important "second": The *Fram*'s successful passage to the northeast around Cape Chelyuskin in the summer of 1893 had proved that the voyage of the *Vega* had not been a matter of luck—that is, the fortunate choice of the one year in many when it could be done—as had been claimed in many quarters after her return.

The aging Baron Nordenskjöld hailed the *Fram*'s triumph with delight. Perhaps now, he thought, my predictions about the usefulness of the Northeast Passage will be recognized as true.

12  *Comrade Stalin's Northern Sea Route*

"Can the voyage of the *Vega* be repeated every year?" Nordenskjöld had been asked many times after his return. The explorer always hedged a bit in answering. He was not sure. But he thought that it could be done in most years, and he certainly looked forward to the early growth of a rich trade between Europe and western Siberia. When he died in 1901, nothing of the sort had happened. Indeed, he was awaiting news of the passage of another vessel in the wakes of the *Lena,* the *Vega,* and the *Fram.*

Arctic commerce, for which the Swedish explorer had seen so great a future, had proved to be unrewarding. European business interests that attempted to establish a regular trade found their profits wiped out by the loss of ships and by seasons of impenetrable ice. Interest soon waned, and there were years when no ships at all sailed to the Ob' and Yenisey rivers.

"The route is really not that difficult," said men who still believed in its promise. Passage would indeed be quite easy if Russia would abandon its attitude of total indifference toward the Arctic, they pointed out. Almost no scientific work had been done up there since Bering's day. There were no aids to navigation and no sailing directions, and the eighteenth-century charts lacked sufficient detail to be of much use in such hazardous waters.

Russia's apathy is understandable. Almost all the industrialists who hoped to realize profits on the country's northern shores were foreigners. Why should she help them to line their pockets?

Czar Alexander III's government took no interest in the Kara Sea Route to Siberia until 1893, when construction of the Trans-Siberian Railway had advanced to the Yenisey River. Then, the builders saw, rather suddenly, that transporting rails to railhead by water would be quicker and cheaper than overland. Russia called in haste upon Captain Joseph Wiggins, a British veteran on the Kara Sea run, to escort four vessels to the river mouth.

With the successful completion of the operation, some of the czar's ministers came to a tardy realization of the Arctic's importance to the country as a whole and urged upon him the development of the northern water route.

Alexander was lukewarm on the subject. At the moment, his nation's internal affairs gave him little time for thought about its far reaches. However, some charting was undertaken in the Kara Sea and at the river mouths. When Nicholas II, the amiable man of slight ability who was to be the last of the czars, succeeded Alexander in 1894, he turned a largely deaf ear to pleas for development of the Arctic route. "We have the railway linking our eastern and western borders now," he reasoned. "What need is there for a northern sea passage?"

That question was answered decisively in 1905, when Russia was at war with Japan. The rail line was jammed with men and military supplies. There was no space at all for food for central Siberia. In April, with a disastrous famine imminent, the czar and his advisers recalled the lesson learned in 1893, and since forgotten. Food must go to the stricken region by way of the Arctic.

The next few weeks saw Russian agents ransacking the ports of northwestern Europe for suitable vessels. Anything tough enough to dare the Kara Sea would do. Price was no object. At length, twenty-two vessels were rounded up—tugs, barges, and steamers—and at summer's end, sixteen of them had managed to deliver twelve thousand tons of food to Siberia. Again it had been proved that the interior could be supplied by water.

In the years following the humiliation by Japan, men who had pleaded for work in the Arctic insisted that Russia need not have suffered defeat in the Pacific, if enough had been known about the entire Northeast Passage. They claimed that the transport of men and materials to the war front would have been eased. Naval reinforcements, moving over the route, might have tipped the scales in the catastrophic Battle of Tsushima Strait, which had ended with the destruction of every Russian ship engaged.

Nicholas was willing to pay attention at last. He authorized the construction of two small icebreakers, the *Taymyr* and the *Vaygach*, which were to explore the entire passage. Launched in 1909, the two vessels started their operations from the Pacific with brief forays into the Arctic in 1910 and 1911. During the next two summers they attempted to pass Cape Chelyuskin, but each time ice conditions turned them back. Then came 1914, a kinder year. They fought their way around the cape before winter caught them in the Kara Sea. The two icebreakers finally sailed into the Atlantic in the summer of 1915, the only vessels to have traversed the whole passage since the *Vega* and the first to make the transit east to west.

In the meantime, a few navigation markers and four telegraph stations had been built at various points on the Kara Sea route. There were plans for others, but the completion of the icebreakers' voyage marked the virtual end of Arctic development by the czarist regime. Russia was faring badly on the battlefields of World War I, which had broken out in 1914, and revolution was brewing. The Arctic must wait for another day.

Czar Nicholas II did not live to see that day, but there may have come a time before his life ended when he regretted his "too little, too late" Arctic policy. In 1917, when he and his family were prisoners of the revolutionists at the old Siberian city of Tobolsk, loyal friends in Moscow devised all sorts of rescue schemes. Apparently, no one ever thought of the water route north through almost unpopulated country to the safety of a neutral vessel in the Arctic. It is possible that none of the rescue planners knew that such a thing might be possible. Nicholas was not a complete fool. Did he ever speculate about that northward-flowing tributary to the Ob' that he could see from his prison?

There was only one summer in which the river could have been used for escape. The revolutionists decided to move the imperial family to Ekaterinburg in the Ural Mountains. There, in July, 1918, Nicholas, his wife, and their five children were murdered en masse.

Russia's new government was quick to reverse the old regime's policy of indifference toward the country's northern resources. Pockets of monarchist resistance were still being crushed in 1919 when the Commission for Study of the North was created. By that time, there were plenty of men in

Moscow who had firsthand information on Siberia—political prisoners, released from long exile, who talked feverishly of resources with which they had become all too familiar. ''There is no doubt that such wealth must be exploited,'' said the commissioners, and advised the construction of ice-breakers, weather-reporting radio stations, and coastal ports.

However, there was still widespread reluctance to accept the idea of the Northeast Passage as a commercially useful route. ''No ship has ever gone through in a single season,'' said critics. ''The *Vega* did not do it, the *Taymayr* and the *Vaygach* failed, and look what happened to the Norwegian, Amundsen, up there.''

A recital of Roald Amundsen's misfortunes on the passage was regarded as a telling argument against spending great sums of money in the Arctic. In 1918, he had set out in his little ship *Maud* to better Nansen's record for a northerly drift across the Arctic. The *Maud* made no such time on the Northeast Passage as the *Vega* and the *Fram* had. She was forced to winter twice and did not reach Bering Strait until 1920. The subsequent two-year drift carried her only as far as the New Siberian Islands, where Amundsen gave up on the project.

Despite the doubts created by the Norwegian's misadventures and the earlier failures to make the passage without wintering, the commissioners' recommendations were adopted. It was becoming increasingly clear that the Soviet Union possessed richer resources in the Far North than had been suspected. New mining discoveries were being made, and reports of oil on Alaska's north slope led to the belief that it must also be present in adjacent areas of Siberia.

The possible value of the passage in time of war had not been forgotten, either, and in this connection, the possession of Wrangel Island, one hundred-odd miles north of Siberia, became of prime importance. At the time, the island was being held by Canadian explorer Vilhjalmur Stefansson for his country, on the grounds that survivors of one of his expeditions had existed there for six months before being rescued. It might well have been claimed by the British, who discovered it in the mid-nineteenth century, or by an American naval expedition which surveyed it in 1881 while searching for the *Jeannette*.

The Russians, who had not set foot on the island until 1911, landed in 1924, raised the hammer-and-sickle flag, and placed Stefansson's colonists under arrest. Wrangel Island was logically a part of Russian territory, they claimed.

In a similar spirit of aggressiveness, the Soviet Union declared that the great seas of the passage—Barents, Kara, Laptev, East Siberian, and Chukchi—were no more than bays on its northern coastline and so should be called "inland sea waters." They were decreed to be off limits for foreign shipping.

During the 1920's, while traffic over the Kara Sea route to western Siberia grew moderately, there was virtually none from the Pacific to the eastern Siberian coast. Josef Stalin, who came to power in 1929, was impatient with the slow progress of Arctic development. Some shaking up was needed in that direction, he declared—a new administration, new faces, and a name for the Northeast Passage that would convey the idea of Russian sovereignty up there.

The result was the creation in 1932 of the Central Administration of the Northern Sea Route, an agency in which Stalin continued to take a very personal interest during his years in power. It was said that he regarded his nation's northern regions as a sort of new American West, and that his fondest dream was to see its potential realized.

Neither money nor effort was to be spared, and no time

was lost in getting the work of the new administration underway. In the spring of 1932, its chief, Arctic veteran O. Yu.
Shmidt, a jovial, bearded giant of a man, went into action.
He readied the icebreaker *Sibiriakov* for a voyage from the
Atlantic to the Pacific, which was to determine whether the
route could indeed be developed for single-season use, as
Nordenskjöld had asserted fifty-three years earlier. Vla-
dimir Voronin, an experienced polar navigator, was named
the ship's captain.

The *Sibiriakov* proved the truth of the Swedish explorer's
claim. Sixty-two days after she sailed from Archangel, she
emerged into the Pacific, the first ship to complete the passage without wintering. None of those who sailed in her
boasted about having been on a pleasure cruise, however. In
the Laptev Sea, battling heavy ice, the ship lost a propeller
blade. Again, near the *Vega*'s wintering place, ice took its
toll—another blade went. Voronin ordered coal and food
thrown overboard, so that the stern could be raised for repairs.

He might as well have saved his time and supplies. Less
than twenty-four hours after he got the ship underway again,
a thrust bearing failed. The following day the propeller shaft
broke, allowing the screw to drop to the bottom. After that,
the captain hoisted sail for a slow reach to the Pacific.

Failure had been a near thing for the *Sibiriakov,* but her
arrival in Bering Strait was hailed as a genuine triumph. The
exultant Shmidt immediately began his planning for the next
summer. In 1933 Comrade Stalin and the world would have
proof that the route was practical for commercial use. All the
doubters would see that cargo could be moved through the
Arctic—passengers, too—Shmidt added.

13 The Loss of the Chelyuskin

Shmidt's ambitious plan to carry cargo and passengers over the whole length of the Northern Sea Route required a specially built ship—one that could deal with ice on her own account, with possibly some help from icebreakers in rounding Cape Chelyuskin and in the eastern seas. He placed orders for the *Chelyuskin*, a 4,000-ton vessel, to be powered by 2,500 horsepower, double that customary in ships of a similar size. For her role as a semi-icebreaker, she needed special frame designing, reinforcement about the bow and forward bulkhead, and compartmenting below the passenger deck for use in case of a forced wintering. In appearance, she was simply a little two-stacker cargo vessel.

The *Chelyuskin* sailed from Murmansk on August 10, 1933, with more than one hundred people aboard—an oddly assorted ship's company which included students, scientists, photographers, newsmen, an artist, a poet, and a few families who were to be landed at the station on Wrangel Island.

For a few of them it was a first time at sea, and many of them had never seen the Arctic. But command rested with the old polar hands, Shmidt and Captain Voronin, and they had a nucleus of the former *Sibiriakov* crew serving under them.

The *Chelyuskin* passengers who eagerly sought a first glimpse of the frozen Arctic were disappointed during the early days of the voyage. Not until the ship had left the Barents Sea behind did she encounter ice. Then, quite suddenly, there was all too much of it for peace of mind in the pilothouse. Voronin was glum, when he toured the ship with Shmidt after his initial attempt at icebreaking. "She's not going to be as good as we thought," he observed, jerking his head at damaged plates and a bend in the stem.

Proceeding with great caution, the *Chelyuskin* reached the cape whose name she bore on September 1. Shmidt called for a conference with the captain of the waiting icebreaker *Krassin*. It was not a comforting session. The *Krassin* had suffered extensive damage that had reduced her icebreaking power by half, said her captain. He doubted that he could help the *Chelyuskin* at all, and added that, although the Laptev and East Siberian Seas should give her little trouble, the Chukchi Sea was another matter. There, the icebreaker *Litke* was struggling to free a number of ships with which she had wintered, and she, too, was in poor condition.

Agreeing with the icebreaker captain that there was no point in insisting upon the *Krassin*'s company, Shmidt and Voronin debated their future course. Should they turn back? Should they attempt to make it alone? "The *Chelyuskin* has cargo for Wrangel Island. If we don't deliver it, no supplies will reach the station until next summer," Shmidt said at last. "I think that we must try."

Summer was drawing to a close as Cape Chelyuskin was left astern, and the need for haste was urgent. With young ice already forming on the leads between floes, Voronin made all possible speed, but the situation deteriorated steadily.

Reconnaisance flights made by the small plane carried on the ship's forward deck revealed no easy passages ahead. The ice was heavy everywhere. The captain was obliged to explore narrowing corridors of water, breaking thin, new crust, occasionally ramming through old floes. He heard the crash and felt the shudder of each assault with mounting apprehension.

In mid-September, Shmidt gave up on reaching Wrangel Island—during that season, at least. Voronin had made it plain enough that they might count themselves fortunate if they got the ship through Bering Strait before winter set in.

The *Chelyuskin* continued her bucking, twisting progress eastward for another week or so. Then, with less than two hundred miles of the voyage remaining, her luck ran out. Caught fast in drifting ice, she was trapped for fourteen days, until an erratic shift in the floes freed her for another few miles of steaming. Near Cape Stone Heart, ice imprisoned her once more. The *Chelyuskin* never moved on her own power again.

The ship's voyaging was far from done, however. For days she moved back and forth with the drift, describing loops within sight of the cape. Her passengers grew sick of Stone Heart and cheered when, at last, the ice began a steady southeasterly movement. On November 3, the ship entered Bering Strait. On November 5, she was halfway through, between Cape Dezhnev and Big Diomede. A plane based on the coast flew a reconnaissance and reported clear water some twelve miles ahead. If the drift held its momentum, freedom and success were a matter of hours distant.

Toward evening, movement ceased for a time. Then, the drift reversed its direction, and with incredible speed, spewed ice field and captive ship back through the strait into the Arctic Ocean.

When the northerly drift continued, Shmidt knew that the

Chelyuskin was in the grip of a powerful current, and was probably doomed. With each mile of northing, the chance of escape grew more remote. He feared that the ship might even be washed into the central polar basin, far beyond the possibility of rescue for her helpless people.

He decided that he must beg for help. The damaged *Litke* lay in Providence Bay just south of the strait. Surely, even at reduced power, she should be able to break a path to the beleaguered *Chelyuskin*.

The response to his plea was prompt and enthusiastic. The *Litke* would sail at once, said her captain. However, his willingness to go on a rescue mission was not shared by his crew. The icebreaker had scarcely weighed anchor when her men began to talk about the danger of being trapped for another eight or nine months. Three days later, the captain informed the *Chelyuskin,* still some twenty-five miles distant, that he was going to accede to demands for a ''ship's soviet'' (a council of crew representatives) to debate the situation.

Shmidt and Voronin were disgusted at this turn of events, for the captain's report on his crew's mood left them in no doubt about the outcome of the ''soviet.'' As head of the Northern Sea Route Administration, Shmidt had the authority to order the icebreaker to proceed in any case, but when he broached this course to members of his staff, the reply was a contemptuous ''Let them go!''

So, the *Litke,* freed of her commitment, sailed away south, and the *Chelyuskin* continued her lonely voyage into the north.

The passengers had always known that they might have to spend a winter in the ice, and they set about preparing for the months of inactivity with courage and resourcefulness. They were ignorant of the fact that Shmidt was readying his command for a disaster that he regarded as probable. The ship had not lived up to his expectations for her. Rivets had been loosened and more plates bent in the Chukchi Sea—he had little doubt that at some point the pressure of the pack would crush her sides.

Selecting five men, he cautioned them to secrecy, lest panic seize the ship, and set them to work in the hold, shifting and organizing supplies, so that they could be salvaged at a moment's notice.

Voronin shared Shmidt's fears for the future. As the weeks passed, with the ship's drift altering from north to northwest, he spent an increasing amount of time in the crow's nest. Later he remembered, "I got to know every single floe. I observed every change in them, and the longer I did this, the clearer it was to me that the *Chelyuskin* would never escape. The dimensions of the pack constantly increased. . . . It was ancient ice—polar pack ice."

For a brief time in December open water appeared around the ship, but the captain saw that there was no means of escape, surrounded as she was. In a few hours, the pack closed in again. It was, indeed, not a motionless world of ice that he surveyed, nor was it silent. Pressure in the pack tumbled floe upon floe—with a grumbling sound like heavy gunfire when the movement was distant, with thunderous sharp reports and shocks throughout the ship when close at hand.

At the outset of the drift, many of the passengers had been a bit unnerved by the noise and the shuddering of the vessel, but they had faith in the *Chelyuskin* and eventually came to accept the commotion with a sort of cool indifference. After the first of the year, there was more concern about saving coal, so that the ship would be able to steam when she was freed, than there was about the hazards of her position.

The end came quite suddenly on February 13 in the midst of a blizzard, with the temperature at minus 22 degrees. Shortly after lunch, one of the engineers ran to Voronin. "A high ridge is coming down and growing fast. It's coming down straight on us!" An anxious crowd lined the port rail on the instant, peering through blinding snow to catch occasional glimpses of a monster ridge bearing down upon the ship. "This is the end!" Voronin told himself.

The first impact came moments later with a crash and the screech of rending metal. Plates bulged and rivets flew like hail. The ship's port side had been gashed along some thirty to forty feet. Water poured into the engine and boiler rooms.

In response to a crackle of orders, Shmidt's long-planned abandonment of the ship began. An unending stream of casks, bundles, and boxes slithered to the ice on planks and ladders that had been flung against the starboard side. The most vital items—food, warm clothing, tents, and sleeping bags—went first, followed by radio equipment, instruments, and the construction materials that had been consigned to Wrangel Island. Each person was given five minutes to gather up his personal belongings. When the final order came, "Everyone on the ice!" the sinking ship had been cleaned of virtually all that might be of use in maintaining life.

Moscow learned of the catastrophic loss of the *Chelyuskin* the following day, and faced the appalling fact that ninety-two men, ten women, and two tiny children must be rescued from the ice ninety miles north of Siberia in the dead of winter. The Russians were equal to the task. Two months later the last of the castaways was airlifted to safety.

"The loss of the *Chelyuskin* is really not significant," the expedition leader asserted during a tumultuous welcome in Moscow. "The Northern Sea Route can be used by cargo vessels—our ship proved it when she entered Bering Strait. The need is for more icebreakers to be stationed along the route to aid in the passage.''

In truth, Shmidt was admitting to having been too ambitious in his planning for 1933 and might well have been censured for an error in judgment when he decided to proceed unescorted from Cape Chelyuskin. The question was never raised.

Indeed, everyone involved in the affair emerged from it in smiling triumph, the survivors and their leader acclaimed for the stamina and courage with which they had set up and maintained a village on the ice while awaiting rescue, the rescuing pilots decorated for pioneering feats of Arctic flying that were as noteworthy as anything that Shmidt had hoped to do with his ship on the Arctic passage.

14  *A Lifeline for the Soviet Union*

The *Chelyuskin* disaster had no effect on the continuing Soviet development of the Northern Sea Route. Improvements came fast during the next few years—new and better icebreakers; such aids to navigation as lighthouses, buoys, and markers; new sailing directions for the whole route; and more accurate ice and weather forecasting.

Only three years after Shmidt lost his ship, the nation was exulting over the fact that 160 vessels had used all or parts of the route during the summer, calling at ports that were mushrooming along the Arctic coast.

Stalin had spoken of developing the resources of the Far North for the benefit of the whole country, but the thought of the route's value in time of war was never far from his mind. Shmidt and his successor, Ivan Papanin, understood this clearly, and in 1939, Papanin claimed with confidence that the 1905 Battle of Tsushima Strait would never be repeated. "In an emergency . . . we shall be able, undisturbed and in a short time, to transfer warships from one sea border of our great Soviet Union to the other," he announced.

Not many months later, Papanin learned that Germany,

Russia's ally for a short time after the outbreak of World War II, had been listening when he made his boast. In the spring of 1940, the Germans asked the Soviet Navy to permit the passage of a merchantman, the *Komet,* offering to pay the costs of icebreaker escort. Friendly relations between the two countries were already crumbling, but for some reason the Russian naval people agreed.

There may have been second thoughts about the decision, however, for when the *Komet* reached the Barents Sea in July, her captain, Robert Eyssen, was immediately suspicious of the arrangements. He was informed that all passes through Novaya Zemlya to the Kara Sea were ice-blocked. His voyage could not begin until August 1, at the earliest.

In constant fear of being discovered by British naval units, he moved his ship from one anchorage to another for a period of four weeks. In the meantime, he was intercepting radio messages from two eastbound convoys which were already sailing over his intended route. Eyssen decided that a deliberate attempt was being made to halt the *Komet*'s voyage, but at last he received authorization to proceed to Matochkin Strait, which separates the northern and southern islands of Novaya Zemlya. An icebreaker would await his arrival, he was told.

There was no icebreaker to be seen as the *Komet* approached the strait. Instead, a boat put out from shore bringing two ice pilots. The German captain had no desire to have a couple of nosy Russians aboard his ship, but the circumstances left him no choice.

There was another three-day delay at the eastern end of the strait, but after that the *Komet* made good time to the vicinity of Cape Chelyuskin, where an icebreaker finally appeared to guide her through Vil'kitskogo Strait. Once the corner had been turned into the Laptev Sea, a second icebreaker took over to forge a way through an ice field. The Soviet vessel steamed off when the ice had been left behind.

Following a course plotted through the New Siberian Islands, Eysson thought that the rest of the trip would be simple, but in the vicinity of the Bear Islands, where Nordenskjöld had seen the first approach of winter, he found trouble that seemed about to spell an end to the *Komet*'s voyage. A third icebreaker joined his ship, bearing an official of the Northern Sea Route administration. He had bad news for the German captain.

"You cannot proceed farther east," the official announced. "I have been ordered to bring you back." He added an excuse about hostile ships in the Bering Sea.

Eyssen protested that he was not worried, but the Russian remained firm. By this time he had informed Moscow that more was at stake than the passage of a small merchantman from the Atlantic to the Pacific. The two ice pilots had had a great deal to say about the ship they had boarded at Matochkin Strait. "She's no merchantman. She's a fully armed raider," they claimed. "There are almost three hundred men aboard her!"

In view of the fact that the Russo-German alliance appeared to be nearing an end, it seemed the height of folly to permit the German ship to escape into the Pacific, where she might soon be preying on Russian commerce. However, Moscow had no real choice in the matter—an icebreaker would be no match for an armed raider if her captain was determined to go on.

At last, there was grudging consent to a continuation of the voyage, and the *Komet* made off. Two days later she passed through Bering Strait after the fastest transit of the Northeast Passage on record—twenty-one and a half days, during which she had been underway only fourteen.

Unhappy as the Russians were about the affair, they knew that the strategic usefulness of the passage had been proved beyond a doubt. And following the abrupt termination of the German alliance not long after the *Komet*'s voyage, it became clear to them that they had a true lifeline in the far North. Mountains of vital supplies and equipment, arranged for under lend-lease agreements with new allies, were being routed to Arctic piers. Eventually, they enabled the Soviet Union to turn back a German invasion and then to mount an offensive.

During the years of World War II, the avalanche of aid never stopped, although in the early part of the conflict, convoys on the Murmansk run suffered stunning losses from German submarine attacks. Later, many of the convoys from the Atlantic were routed farther north, passing above Novaya Zemlya, with icebreaker assistance, to leave their cargoes at Kara Sea ports.

Aid to Russia came from the other end of the passage, too. Between 1942 and 1945, 120 voyages were made from the Pacific coast of North America to new cities on a coast which had numbered no more than a handful of native setlements only a few years earlier. Fifty-four vessels forged through the Chuckchi, East Siberian, and Laptev Seas all the way to Tiksi, a seaport built on the shores of the Lena Delta not far from where the *Jeannette*'s Chief Melville had landed with his starving men in 1881.

When peace came in 1945, the usefulness of the Northern Sea Route had been demonstrated in magnificent style. Not only could the Soviet Union supply her vast northeastern territory and carry away its wealth by water, but Papanin's claim that there would never be another Tsushima was known to be the truth.

15 The Soviet Ice May Yet Be Thawed

During the years of the lend-lease voyages, American ships calling at eastern Siberian ports from the Pacific were crewed by Russians (in order to maintain the Soviet Union's neutrality with Japan), but this was not the case in the western Arctic. Hundreds of foreign seamen glimpsed bustling towns and cities along that coast, and marveled at the sight of scores of convoys passing over a waterway so remote that few of them had been aware of its existence.

If they thought to see it again in time of peace, however, they were mistaken. When the Iron Curtain came down around Russia and its European satellites after the war's end, it fell in the Arctic too. Strangers were no longer welcome, now that the need for them had ended.

Undisturbed by prying eyes, Russia plunged anew into the development of her Far North. In the years since, there has been scant news from that part of the world. The little that is known about it has been largely gleaned from reports made public for the benefit of Soviet citizens—reports that tell of astonishing progress.

Guided by extremely accurate weather and ice forecasting,

and convoyed by the ships of an icebreaker fleet, hundreds of vessels move for almost half the year through a passage that Nordenskjöld thought might be open for a few weeks. Some vessels make two and three round trips a year, and one voyage from Murmansk to Bering Strait was logged in an amazing ten days.

The Russians have talked seriously of keeping the entire passage open throughout the year, and since 1970 they have proved that the western portion from the Atlantic to the Yenisey can be used in all seasons. Each year a midwinter convoy, escorted by several icebreakers, makes a round trip over the route. Apparently, there have been no mishaps, although conditions in the Kara Sea are often rugged enough to keep the merchant crews on edge.

One reason for Soviet optimism about a year-round passage may be their nuclear-powered icebreaker *Arktika*. In the spring of 1975 she left the shipyard to join the Arctic fleet which since 1960 has included the atomic icebreaker *Lenin*. The *Arktika* is said to be twice as powerful as the 44,000-horsepower *Lenin*. In August 1977, the *Arktika* became the first surface ship to reach the North Pole, thus realizing the dream of the intrepid explorers of many countries.

The Soviet Union has not been able to resist a comparison of the Northern Sea Route with the more famous Northwest Passage: In most years no vessels at all sail through the Arctic north of Canada, and only one large ship, the supertanker *Manhattan,* on an exploratory voyage in 1969, has transited the Northwest Passage.

Authorities on maritime transportation claim that Russia's Arctic route has not proved to be the commercial success that early developers foresaw. They cite their belief that few, if any, foreign vessels have yet availed themselves of a 1967 offer to use the passage.

The Soviet Union probably does not care, for Siberia is proving to be a treasure house for which maintenance of the route is imperative. Remote river basins are spewing forth incredible torrents of the earth's wealth. In fact, it is believed that the world's greatest reserves of such things as gold, diamonds, coal, lead, nickel, iron ore, manganese, and cobalt may lie beneath Siberian permafrost.

Scientific reindeer husbandry has increased domestic herds to millions of animals, providing a meat supply for local consumption and for export. The sealing and fishing industries have grown tremendously, and timber remains a prime export.

An energy-hungry world listens with most interest, however, to news from Siberian oilfields. Until the early 1960's, about four fifths of the Soviet Union's major crude-oil resources came from a huge field reaching from the middle Volga area to the Ural Mountains. Another rich field—one of the world's oldest—lay in the area of the Caspian Sea. Russia's far east was served by wells on the northern tip of Sakhalin Island off the Pacific shore. Then geologists exploring the great "wet Northwest" area of Siberia made what one observer has termed the "greatest single oil strike in the history of geology." Well after well drilled on the desolate, unending plain of the Ob' River basin came in with enormous quantities of high-grade crude oil. Huge supplies of natural gas were found there too. The next three years saw the discovery of oil all across Siberia, giving rise to speculation that the Sakhalin Island field may lie at the edge of one great reservoir.

Oil- and gas-pipeline construction has gone ahead with furious speed. The longest line in the world—2,600 miles—now delivers natural gas to Moscow and Leningrad. And rails are being laid as fast as possible to connect the

North with the Trans-Siberian line. But the Northern Sea Route is needed as never before.

Heavy equipment for oilfields and mines comes by water, as do many of the craft—tugs, barges, passenger hydrofoils, and floating cranes—that are needed on Siberia's 200,000 navigable river miles. Most supplies for the immense republic of Yakutsk are delivered by way of the Arctic and the Lena River, and its exports go out over the same route. There are towns in eastern Siberia, served by neither road nor rail, that could not survive without seaborne commerce. The Soviets cannot afford to neglect their northern passage.

The continuing concern over its development is reflected in the tenth Five-Year Plan for the years 1976 through 1980. Regarding the Northern Sea Route, the plan calls for increased freight turnover on the route, Arctic port improvement, and measures aimed at lengthening the navigation period.

The development of Siberian resources would not be possible without people, of course—hordes of courageous, energetic men and women with a pioneering spirit. The government made its first appeals for Arctic colonists early in the program, and Russians, particularly young people, responded joyously to the challenge of conquering a new world. Over the years they have come by the thousands to build the things that have made civilization in the Arctic possible—dams, roads, factories, refineries, cities.

Visitors from the West who have lately traveled in Siberia have noted this tremendous development with awe. In particular, the new cities are sources of amazement. Populations of over 100,000 are not uncommon. Norilsk, founded in 1936, has 136,000 residents, and Yakutsk, a very old city, has 108,000. Multistoried masonry structures, shops, theaters, restaurants, and hospitals have risen on the permafrost

of one of earth's coldest regions, and more elaborate construction is going on constantly.

The smaller centers of population enjoy the comforts and
pleasures of civilization too. Electricity, running water, central heating, community centers, and sports stadia are com-
monplace. The one thousand or so people of Kolymskaya, on
the Kolyma River not far from the Arctic shore, will eventually live and carry on their business in a complex of buildings
laced together with aluminum and Plexiglas corridors.

"But what will happen to these places when the resources
of their areas are exhausted?" ask skeptical tourists, remem-
bering boom towns in the American West that sprang up and
died within the space of a few years.

"We want to bring people to the Arctic on a permanent
basis," Soviet officials reply, "and we are building for that.
Conservation has always been a part of our planning. Siberia
will not be stripped of its wealth and abandoned. The cities
and towns will continue to grow."

Once, during the postwar period of miracle working in the
Arctic, Russia drew stunned attention to her polar regions.
On a day late in October, 1961, sensitive instruments at
weather stations throughout the world recorded a sudden
change in atmospheric pressure very like that produced by a
severe electrical storm. Meteorologists were startled. What
on earth or in the skies above it could have happened?

The answer was in the news the following day. A ther-
monuclear device of tremendous power had been fired
somewhere north of Russia. The Soviet Union confessed
with little delay to having ended a three-year moratorium on
such testing with the biggest bang ever heard on earth,
touched off at a new testing ground on Novaya Zemlya.

The announcement created a globe-encircling shock wave
of dismay, followed by intense interest in a part of the world

of which little was generally known. Map study revealed adequate information on the land masses, but marine charts were another thing. They showed most of the waters of the Soviet Arctic as an uncharted blank, for Russia had never made the results of her hydrographic surveys available to other nations.

By coincidence, the United States was at the moment preparing to complete a study of the entire Arctic Basin by surveying those unknown waters—insofar as the Russians would permit.

In the summer of 1962, the icebreakers *Burton Island* and *Northwind* worked in the Chukchi Sea without incident, and the next year the *Northwind* edged her way across the East Siberian and Laptev Seas to within a few miles of Cape Chelyuskin, the first vessel to carry the American ensign deep into the Soviet Arctic. Strangely, it may not have been a first time in those waters for the ship. As a lend-lease icebreaker during World War II, called by the Russians *Severnyy Veter (North Wind),* she could well have been employed by them on the Northern Sea Route.

The *Burton Island* carried on the survey in 1964, working in the East Siberian Sea. Her people had a hair-raising experience during that cruise. While she was in the strait south of Wrangel Island, a huge convoy appeared, en route to Bering Strait. Led by Soviet icebreakers, it was made up of both merchantmen and naval vessels. The *Burton Island*'s captain, a "wild man," according to his crew, decided to take a closer look. He joined the convoy, weaving in and out while camera shutters clicked madly.

Then he ordered a helicopter aloft. Promptly, the Russians sent one up to play a deadly game. It took station over the interloper, forcing it down steadily toward the water, and each time the American "chopper" escaped, the more maneuverable Russian machine topped it, and again began to blast it down. "It looked to me as though he were trying to put his wheels in my rotor blades," said the shaken pilot when he landed.

The captain was obliged to give up on helicopter flights, but he was well satisfied with his inspection, when at last he turned away.

No doubt the men who were already planning the *Northwind*'s attempted 1965 transit of the Northeast Passage read the *Burton Island* report with a great deal of interest. There might well be a chilly reception for the former *Severnyy Veter* when she entered the Kara Sea, and if that was the case (as it proved to be), it was decided that Captain Ayers would receive a prompt change of orders. The United States was not prepared to upset the delicate balance of Soviet-American relations with an Arctic "incident."

Less than two years after Ayers retreated with the *Northwind*, Russia appeared to be undergoing a change of heart regarding her interocean passage. Early in the spring of 1967, she announced that henceforth the Northern Sea Route would be open to foreign traffic, subject to thirty days' notice and the payment of tolls. A brochure was issued detailing its advantages, and a voyage was staged to prove them. That summer a Soviet freighter loaded cargo in Le Havre, Antwerp, Rotterdam, and Hamburg. Twenty-seven days after sailing from Hamburg, she docked in Yokohama. It was a spectacular demonstration, and by accident, it had been timed perfectly. War in the Middle East had just resulted in the closure of the Suez Canal.

Shippers conceded the Northern Sea Route's advantages, but there was no rush to file transit applications with the Russians. Pilotage fees and tolls were steep enough to cause more than a little doubt about the economy of using it.

"Perhaps sometime. Not now," said a Norwegian shipping man, expressing the universal concern. "The expense, together with possible delays, could cancel all the time advantage."

However, Russia's relaxed attitude toward the Arctic passage was cheering news in many quarters. Probably it was greeted with more enthusiasm aboard a pair of United States Coast Guard icebreakers than anywhere else.

Months earlier, the *Edisto*'s captain, William K. Earle, had proposed for the summer of 1967 an ambitious project that called for the transit of both the Northeast and Northwest Passages. He had requested permission to sail northeast from Europe with two ships, his own and the *Eastwind*. His planned route would take the squadron north of Novaya Zemlya, Severnaya Zemlya, and the New Siberian Islands, thus avoiding straits where the Russians might turn it back. After crossing the Chukchi and Beaufort Seas, the icebreakers were to transit the Northwest Passage and return to the Atlantic by way of Baffin Bay. They would then have completed the first single-season circumnavigation of the Arctic Basin.

Captain Earle had alternatives to the plan. If the ice in the North proved to be too heavy, he would try for passage through Vil'kitskogo Strait and Sannikov Strait, the latter in the New Siberian Islands. In the light of Russia's change of policy on the Northern Sea Route, it seemed that those alternatives might be possible.

Permission for the voyage was granted, and the two ships sailed, with everyone aboard optimistic about an uneventful summer. They had not been long in the Arctic before that notion disappeared. Soviet bombers were keeping them under constant surveillance. Clearly, American "warships" (so Russian officials termed the icebreakers) would still not be permitted to cruise those waters unescorted.

There were no threatening gestures such as the *Northwind* had experienced, however, and all went well until the squadron approached Severnaya Zemlya. At that point, it became evident that ice conditions were much worse than they had been in 1965. Impenetrable ice lay over the area where the *Northwind* had sailed in open water two years before. For five days Captain Earle groped for an opening that would lead him over the top of Severnaya Zemlya. Then, since he could waste no more time, he decided to resort to the first of his alternatives.

The squadron sailed for Vil'kitskogo Strait, and the *Edisto* radioed a message to Port Dickson, communications center for the route:

THIS SQUADRON WILL, ON OR ABOUT 31 AUGUST 1967, MAKE A PEACEFUL AND INNOCENT PASSAGE THROUGH THE STRAITS . . . , ADHERING TO THE CENTERLINE AS CLOSELY AS POSSIBLE, AND MAKING NO DEVIATION OR DELAY.

"IS YU AMERICAN ICEBREAKER?" Radio Dickson queried before sending the message on to Moscow.

The term "innocent passage" in Captain Earle's message had a meaning that is not immediately evident. It is the one which describes the right of vessels to pass through the territorial waters of any nation en route from one port to another. Earle had not asked for this right. He had stated his intention of exercising it—perhaps a bit of bluff. If so, it was useless. Soviet diplomats went to work within an hour of hearing from Port Dickson.

The following morning orders came from the commandant, United States Coast Guard, canceling the remainder of the voyage. Again an American transit of the Northeast Passage had been refused.

Bitterly disappointed, the men aboard the two ships still found some comfort in the situation. After all, they had heard no such words of challenge as the *Northwind* had, and that was a promising development. Was it possible, they wondered, that the Soviet Union might be willing to yield ''innocent passage'' through her straits when another American ship appeared in the Arctic north of Siberia?

''This frozen portion of the Iron Curtain may yet be thawed,'' one man observed, as the *Edisto* and the *Eastwind* turned westward.

There has since been no effort to test such a hypothetical ''thawing,'' but there is a chance that one day a real thaw may occur in the Siberian Arctic. For years Soviet scientists and engineers have talked of reversing the flow of the Pechora, Ob', and Yenisey Rivers, in order to water land in the arid South and raise the level of the Caspian Sea, which has been dropping steadily. Apparently, some planning has been done for a years-long project.

Climatologists regard the proposal with mingled awe and horror. They foresee changes in climate over a large portion of the globe and claim that the vast reduction of fresh water pouring into the Arctic could not fail to result in thinner ice along the coast—perhaps no ice at all.

Would the Northeast Passage then become the interocean highway of which men first dreamed more than four hundred years ago? Or would road, rail, and air transport have so developed by that time that its existence, free or frozen, would be a matter of no consequence?

The world's calendars will have marked a new century before there are answers to the questions about the future of the Northeast Passage.

Bibliography

Bibliography

Andreyev, A. I., editor, *Russian Discoveries in the Pacific and in North America in the Eighteenth and Nineteenth Centuries*. A collection of materials. Ann Arbor, Mich.: Published for the American Council of Learned Societies by J. W. Edwards, 1952.

Armstrong, Terence, *The Northern Sea Route*. London: Cambridge University Press, 1952.

——— *Soviet Northern Development, with Some Alaskan Parallels and Contrasts*. Fairbanks: Institute of Social, Economic and Government Research, University of Alaska, 1970.

De Long, J. K. J., *The Barents Relics*. London: Trubner & Company, 1877.

Edwards, Deltus M., *The Toll of the Arctic Seas*. New York: Henry Holt and Company, 1910.

Ellsberg, Edward, *Hell On Ice*. New York: Dodd Mead, 1938.

Fairhall, David, *Russian Sea Power*. Boston: Gambit, Inc., 1971.

Ford, Corey, *Where the Sea Breaks Its Back*. Boston: Little, Brown and Company, 1966.

Gelder, F. A., *Bering's Voyages*. New York: American Geographical Society, 1922.

Goodhue, Cornelia, *Journey into the Fog*. Garden City, N.Y.: Doubleday, Doran & Company, 1944.

Gruber, Ruth, *I Went to the Soviet Arctic*. New York: Simon and Schuster, 1939.

Hakluyt, Richard, *Hakluyt's Voyages—Selections*. Edited by Irwin R. Baker. New York: The Viking Press, 1965.

Janvier, Thomas A., *Henry Hudson*. New York: Harper & Brothers, 1909.

Lensen, George A., *Russia's Eastward Expansion*. Englewood Cliffs, N.J.: Prentice-Hall, Inc., 1964.

Massie, Robert K., *Nicholas and Alexandra*. New York: Atheneum Publishers, 1967.

Members of the Expedition, *The Voyage of the Chelyuskin*. New York: The Macmillan Company, 1935.

Mirsky, Jeannette, *To the Arctic!* New York: Alfred A. Knopf, 1948.

Mowat, Farley, *The Siberians*. Boston: Little, Brown, 1970.

Nansen, Fridtjof, *Farthest North*. New York: Harper & Brothers, 1897.

———*In Northern Mists*. New York: Frederick A. Stokes Company, 1911.

Nordenskjöld, A. E., *The Voyage of the Vega*. London: The Macmillan Company, 1885.

Parry, J. H., *The Age of Reconnaissance*. New York: The World Publishing Company, 1963.

Payer, Julius, *New Lands Within the Arctic Circle*. New York: D. Appleton and Company, 1877.

Penrose, Boies, *Travel and Discovery in the Renaissance*. Cambridge, Mass.: Harvard University Press, 1952.

Petrow, Richard, *Across the Top of Russia*. New York: D. McKay, 1967.

St. George, George, *Siberia, the New Frontier*. New York: David McKay Company, Inc., 1969.

Smolka, H. P., *40,000 Against the Arctic*. New York: William Morrow & Company, 1937.

Stefansson, Vilhjalmur, *Great Adventures and Explorations*. New York: The Dial Press, 1947.

——— *Northwest to Fortune*. New York: Duell, Sloan and Pearce, 1918.

Periodicals

"Arctic Tragedies of the Past," *Outlook,* October 2, 1909.

Armstrong, T. E., "Place-Names in the Soviet Arctic," *The Polar Record,* January 1950.

——— "Soviet Drift in the G. Sedov," *The Polar Record,* December, 1947.

——— "The Soviet North in the Tenth Five-year Plan, 1967–80," *The Polar Record,* May, 1976.

——— "The Voyage of the *Komet* Along the Northern Sea Route," *The Polar Record,* May, 1949.

"The Baldwin–Ziegler Expedition to Franz Josef Land," *Scientific American,* May 4, 1901.

Bernstein, Leon, "The *Chelyuskin* Epic," *Living Age,* September, 1934.

Brice, Arthur M., "Nearest Village to the North Pole," *Outlook,* December, 1899.

"Captain Brusilov's Arctic Expedition," *Scientific American,* February 27, 1915.

Curry, Henry B., "The First True Polar Voyage," *Asia,* August, 1926.

Dickie, Francis, "The 'Booming' Arctic," *Living Age,* September 15, 1927.

"Hunting for Oil Beneath Polar Snows," *Popular Mechanics,* October, 1924.

Laforest, Captain T. J., "Strategic Significance of the Northern Sea Route," *United States Naval Institute Proceedings,* December, 1967.

Le Bourdais, D. M. "Staking Wrangel Island," *Asia,* April, 1925.

Lee, Sidney, "Arctic Exploration in Shakespeare's Era," *Living Age,* April, 1913.

McConnell, B. M., "Rescue of the *Karluk* Survivors," *Harper's Monthly,* February, 1915.

Melville, Rear Admiral George Wallace, "The Baltic Fleet and the Northeast Passage," *North American Review,* August, 1904.

Mirza, Princess Nusrat Ali, "Potential Wealth of the Arctic," *Nineteenth Century,* October, 1923.

"New Distance Record in Soviet's Arctic Conquest," *Newsweek,* August 1, 1936.

Peckham, Richard, "Pioneers of Arctic Colonization," *Travel,* June, 1925.

"Piercing Arctic, etc." *Literary Digest,* May 2, 1936.

Powys, Llewelyn, "An Arctic Wintering," *Century,* February, 1929.

"Resumé of the Fiala–Ziegler Expedition," *Scientific American,* August 19, 1905.

"Russian Northern Route Not Used," *United States Naval Institute Proceedings,* April, 1968.

"Russia's Arctic Expedition," *Scientific American,* August 1, 1908.

"Soviet Union Launches Second Nuclear-Powered Icebreaker,'' *United States Naval Institute Proceedings,* January, 1974.

''Strange Doings in the Arctic Circle,'' *Current Opinion,* December, 1924.

Synhorst, Captain Gerald E., ''Soviet Strategic Interest in the Maritime Arctic,'' *United States Naval Institute Proceedings,* May, 1973.

''To Siberia by Sea,'' *Harper's Weekly,* December 21, 1912.

''The Tragic Crusoes of Wrangel Island,'' *Literary Digest,* December 8, 1923.

Wells, Robert D., ''The Icy 'Neyt'!'' *United States Naval Institute Proceedings,* April, 1968.

''What Have the Russians Discovered in the Arctic?'' *Scientific American,* October 25, 1913.

Yarbon, E. R., ''Russia's Northern Seaway,'' *Nineteenth Century,* January, 1944.

Index

Index